D1088150

CERAMICS SOURCE BOOK

ERROL MANNERS

A QUARTO BOOK

First published in Great Britain in 1990
by Collins & Brown Limited
Mercury House
195 Knightsbridge
London SW7 1RE

A CIP catalogue record for this book is available from the
British Library.

ISBN 1 85585 020 6

This book was designed and produced by
Quarto Publishing plc
The Old Brewery
6 Blundell Street
London N7 9BH

Senior Editor Cathy Meeus
Editor Judith Warren

Art Director Moira Clinch
Designer Karin Skånberg
Picture Researcher Anne-Marie Ehrlich

Typeset by Ampersand Typesetting (Bournemouth) Ltd
Manufactured in Hong Kong by Regent Publishing Services Ltd
Printed in Hong Kong by Leefung Asco Printers Ltd

Northern Song celadon box and cover, 11th – 12th century.

CONTENTS

FOREWORD

From very simple basic elements – clay, pigment and glaze – over the centuries man has created a vast range of ceramics. To achieve this astounding variety many differing types of material and firing methods have been used. Today all phases of such production can be minutely controlled guaranteeing uniform and predictable results.

How the craft got to this point is a long and fascinating story played out over many centuries in the Far East with the European potters joining in only at a relatively late stage. It is a story of the adaptation of ideas from other disciplines initially from metalwork, but later drawing on the work of the sculptor, the painter, the engraver and even the fabric designer. Potters have also taken their inspiration from the things and activities around them, the world of the court, the sports field, the theatre and the whole sphere of myth and legend. Above all they have sought inspiration in their own tradition. For the history of ceramics is a story of borrowing: borrowing ideas, borrowing techniques and borrowing designs.

The classic example of this cross fertilization of ideas between ceramic traditions, is the extraordinary reaction of the European porcelain manufacturers to the influence of the Kakiemon wares of Japan. These were to provide an international source of inspiration. European products in the

Kakiemon style, which started out as pure imitation, swiftly assumed a character of their own adapting the shapes and designs of the Japanese potters to purposes quite alien to their original conception.

The majority of artefacts suffer physical changes with the passage of time: paintings darken and crackle, and furniture changes colour and tends to disintegrate if neglected. One of the extraordinary characteristics of the potter's product is that, unless it is damaged or exposed to unfavourable conditions, it remains exactly the same as when it left the kiln. Thus the early Ming blue and white dish looks exactly the same today as it did 600 years ago.

So while many pieces of porcelain are admired because they are old, they are in fact "new" in terms of condition and enable us to see them just as their creators did. This provides an unparalleled insight into the taste of past ages and the exact colours and shapes that they found pleasing.

To appreciate fully the evolution of the potter's craft, it is necessary to understand why and how they did what they did and from where they drew their inspiration and ideas. I am sure that this book will help to provide answers to these questions.

A Medici porcelain triple flask. The earliest European porcelain, made in Florence between 1575 and 1587, drew on Urbino maiolica for its shapes and Chinese porcelain for its decorative inspiration.

Hugo Morley-Fletcher

Hugo Morley-Fletcher

INTRODUCTION

From prehistoric times to the present day the unique properties of clay have served mankind. As a medium its adaptability is enormous, but above all its versatility has been exploited most fully in the form of vessels. The primary function of these vessels has been practical, for containing food and liquids, but the makers' urge to decorate them is evident from the earliest periods. And whilst the greater part of the production has been utilitarian, a proportion has, wittingly or unwittingly, achieved artistic distinction.

The aesthetic scope of ceramics is as wide as art itself; on the one hand there is the obvious high quality of the finest porcelains and on the other, great beauty can be found in the handsome proportions of a well-potted rustic jug. Whilst personal preference will dictate which groups of ceramics appeal most strongly, it is important to assess a work in its own context. To judge an austere Song dynasty celadon by the richness of its decoration or an elaborate Renaissance maiolica dish by the refinement of its potting would be to misunderstand the intention of the work.

The interest and importance of much pottery lies not only in its aesthetic qualities but in the light it can throw on both the history and sensibilities of the societies in which it was made. The habit of burying the dead with a range of treasures and useful items was widespread in earlier cultures and has left an enormous legacy of ceramics which, depending on the soil conditions, have survived in an excellent state of preservation. For some of the earliest societies pottery may be the most illuminating indicator of their customs, wealth and trading patterns. In more sophisticated societies like ancient Greece and Tang China, painted and sculptural tomb pottery gives an extraordinarily detailed picture of many aspects of life which would never have been recorded in their literature. The ceramics of any society reflect its prevailing spirit and one can detect the emerging confidence of Renaissance Italy in the development of its maiolica, or the extreme refinement of the Imperial court of Song China before it fell beneath the dynamic onslaught of the Yuan Mongols. The finest wares were inevitably made for the wealthiest sections of society, usually the aristocracy or the merchant classes, but rustic and provincial potters frequently produced work of great distinction, such as the late 17th century English slipware dishes, which many people today find more sympathetic than the refined early porcelains.

Throughout the history of ceramics until relatively recently, potters have generally made efforts to refine their materials and techniques. It was only in Japan that some potters sought to recapture the more rugged qualities and imperfections of primitive wares. Many modern potters have now reacted against the lifeless mechanical perfection of industrial ceramics, and have revelled in hand-made qualities of pottery.

A Greek black-figure Hydria, Athens, 550 BC. The Greek potters produced their greatest work in Athens in the 5th and 6th centuries BC. The highly refined body was burnished to a high gloss and the lack of glaze accentuated the angularity and clarity of form. Figure painting was the favoured form of decoration and it affords a remarkable record of Greek mythology and contemporary life.

A maiolica dish, Faenza, c. 1525, showing Joseph discovering the cup that he had put in the sack. The potters of the Italian Renaissance created a palette of high temperature colours which allowed them to mirror the achievements of contemporary art. A unique property of fired pottery is that it allows us to see Renaissance painting in its unfaded and vivid original colours.

Earthenware, stoneware and porcelain

Ceramic bodies fall roughly into three major groups fired at increasingly high temperatures: earthenware, stoneware and porcelain. The Chinese, however, unlike Western potters, do not make clear distinctions between these groups and there are many hybrid wares that do not fall neatly into these categories. The term "pottery" is commonly used to describe earthenware and stoneware.

Earthenware, fired up to about 1200°C, is the simplest ceramic body and is common to all cultures. Its main shortcoming is that it is porous to water, although this can have its use in that evaporation of water from the surface will cool the contents of a vessel. Earthenware can be brought to a considerable degree of refinement by levigating the clay to remove larger particles, and by covering it in a slip of fine liquid clay which may be burnished by rubbing with a stone or bone to quite a high gloss. The ancient Greek and the pre-Columbian wares are amongst the finest examples of this type of pottery.

Earthenware is often given a vitreous glaze that is completely non-porous. Early glazes were usually made from silica to which lead oxide was added to reduce its melting point; this forms a translucent glassy skin over the vessel which remains a quite separate layer and can be coloured with metal oxides. Notable amongst lead-glazed wares are those of Tang China and European slip and sgraffito wares. The addition of tin oxide opacifies the glaze so that it forms a brilliant white surface which is an excellent background for painted decoration. Tin-glazed ware, in the Middle East, became the most important form

of European pottery until the discovery of porcelain in the 18th century.

Stoneware is fired to a temperature of about 1250°C. It is very hard and durable and almost totally impervious to liquids, even when unglazed. The stoneware tradition of China produced the finest range of glazes ever developed and in the Song dynasty arguably some of the finest pots ever made. In Germany and England the salt glaze technique, by which a handful of salt thrown into the kiln reacts with the surface to form a thin, tight-fitting glaze, was widely used.

Hard-paste or "true" porcelain is the highest technical achievement of the potter. It is formed from china clay and china stone, known to the Chinese as kaolin and petuntse, and fired to 1300-1400°C. At its best it is brilliant white, translucent and very hard, with a tight-fitting glaze. Porcelain was discovered in China in the 7th or 8th century AD, brought to its full maturity in the 14th century – and was the envy of the world. The technique of porcelain manufacture passed to Korea and by the 17th century to Japan; it was independently rediscovered at Meissen in the early 18th century and has since remained the mainstay of the ceramic industry in Europe and the East.

The interior of a warehouse for the East India market, c. 1770–1820. The trade between East and West has continued since ancient times. Its great blossoming in the 17th and 18th centuries brought about a rapid cross fertilization of ideas which radically affected the ceramics of both Europe and Asia.

Attempts to imitate true porcelain in 18th century Europe gave rise to soft-paste porcelains of varying compositions, which are fired at a lower temperature than true porcelains. The best soft-paste porcelains, such as Vincennes, can rival and even surpass true porcelain in beauty but they generally lack its strength. The 19th century saw the creation of bone china, still much used in England, and various hybrid porcelains.

Semi-vitreous glazed quartz fritwares were discovered in ancient Egypt; these were refined and became widespread in the finer Islamic pottery from the 12th century

A Ming style blue and white vase, Yongzheng period, 1723–1735. The classic Ming porcelains of the 15th century set the standard by which all later blue and white porcelains must be judged. In China attempts to copy them came closest to success in the 18th century but in spite of the fine quality of this example the spirit of Ming painting was never quite matched.

Meissen figure of a bear, 1732–35. The almost life-size series of animals ordered by Augustus the Strong for his Japanese Palace at Dresden are amongst the greatest achievements of early European porcelain. But the problems of firing such a large mass of porcelain were only partly conquered.

onwards. Fritwares can be slightly translucent when finely potted and their alkaline nature allows for some brilliant glaze colours (notably a turquoise derived from copper) which are unattainable on other wares. Their lack of strength is the chief defect and consequently it is rare today to find the earlier wares in good condition. Fritware can be considered a type of soft-paste porcelain.

In 18th century England a number of new wares were developed, which can be described either as high-fired earthenwares or stonewares. The most important was creamware, which lent itself well to mass production and was instrumental in leading to the rise of the Staffordshire potteries. Amongst the other notable wares were Wedgwood's jasperware and the strong ironstone china.

Form and decoration

Ceramic wares can be formed in a variety of ways and many of the earliest techniques are still widely used. Pots may be built by coiling strands of clay and smoothing the joints, or they may be moulded or formed by luting together slabs of clay. Slip-casting, in which a liquid suspension of clay is poured into a plaster of Paris mould, allows for intricate forms such as figure models to be made, and the moulds can be re-used. Above all, the wheel

A Sèvres "Saigon" vase, 1895, decorated in a cloisonné technique. By the 19th century the technical excellence of ceramics had become commonplace but the search for fresh inspiration stimulated a great diversity of styles that has continued unabated throughout the 20th century.

has been the potter's most useful tool; the action of the centrifugal force allows the potter to create vessels quickly and fluently with the minimum effort. Pots can also be turned on the wheel with a blade when they are "leather-hard" to reduce their thickness, sharpen features, and create detail. The wheel has dictated the characteristic circular cross-section of much pottery.

Potters have at their disposal a wide range of decorative techniques. The clay body can be incised or carved, as was perfected by the Chinese and Islamic potters; slip can be trailed or painted onto the body. Trial and error has revealed the colouring possibilities of the many minerals which can withstand the firing temperatures. The subtle colour gradations of the maiolica from Faenza or the brilliant enamels of the Kakiemon and famille verte palettes of Japan and China are impressive achievements by any standards.

One attribute of ceramics sets it apart from the other fine and decorative arts, and that is the ability of ceramic ware to remain unchanged in surface and colour. Time gives a patination to paintings, furniture and metalwork, mellowing the colours; we are not accustomed to seeing Old Master paintings in their original brilliance. But when one looks at well-preserved ceramics one can see them exactly as they were intended to look and, in that sense, they are a unique window on the past.

The influence of trading links

With the exception of the pre-Columbian wares of America, most of the world's ceramic traditions are to some degree interrelated. Although there were long periods of cultural isolation, outside influences were frequently an important stimulation to potters. Ceramics, like textiles, seem to cross cultural frontiers more readily than the fine arts. Whereas the content and purpose of paintings from alien cultures might be incomprehensible, the common language of ceramic forms and functions makes them more readily approachable. Dishes, bowls, vases and ewers are common to all but the most primitive societies; if the proportions and decoration of a foreign ewer might be strange, its function would be clear and if it proved superior to the local wares it would be imitated.

A hand-built burnished form by Magdalene Odundo, c. 1985. The contemporary studio potters have drawn on many traditions across the world and there are more consciously creative potters than ever before. This finely burnished vase evokes the natural forms of gourds and fruit, combined with an angularity and surface quality that recalls the vases of ancient Greece.

From the earliest times there was trade between the ancient Mediterranean civilizations and China, along the caravan routes of the Middle East. It is possible to trace the Greek acanthus leaf ornament through the Hellenistic architecture of the Near East to the lotus scrolls of Central Asian Buddhist temples. The lotus scroll also appears on the ceramics of China, and returned to the West through the Mongol conquest of Persia and Turkey.

Of all the world's ceramic traditions, that of China has been the most dominant and influential, both technically and artistically. China considered other countries to be, in varying degrees, barbaric, but in spite of this it was open to outside influences. Many Chinese forms show a debt to Islamic metalware; both Buddhists and the invading Mongols introduced a rich vocabulary of ornament, and in the 18th century the enamels of the famille rose palette were introduced from Europe. However, China primarily exported its accomplishments – China's technical advances such as its celadon glazes and porcelain, were centuries ahead of the rest of the world. It was the first country, since the ancient civilizations, to raise ceramics to a highly regarded art, notably in the Tang and Song dynasties. In the East, China's influence dominated the fine 12th and 13th century wares of Korea and the later Korean potters were themselves a decisive influence in Japan. In south-east Asia many of the finest wares of Vietnam and Thailand were copies of Chinese celadons and blue and white porcelains.

In spite of having considerable trading links with China the potters of the Middle East did not adopt the Chinese stoneware and porcelain technologies, but Chinese imports encouraged them to refine their own ceramic materials, such as tin glazed earthenware and fritwares. Handsome copies of Chinese monochromes and blue and white wares formed an important part of the output of Islamic potters. But the most interesting Islamic wares, such as the lustre wares and the brilliantly designed Iznik wares were derived from their own rich and constantly developing artistic tradition. The pottery of Europe in its turn owes a great debt to Islam. The technology for tin glaze earthenware was imported from the Middle East, as was the lustre technique.

The Northampton Amphora, c. 540 BC.

CHAPTER · ONE

THE ANCIENT WORLD

"O Attic shape! Fair Attitude! with brede
Of marble men and maidens overwrought."

JOHN KEATS – ODE ON A GRECIAN URN

THE ANCIENT WORLD

Samarra ware bowl, Northern Mesopotamia, 5000–4500 BC. The finely potted Samarra wares were painted on a cream slip with carefully arranged geometric patterns or occasional human and animal figures. The early Mesopotamian settlements saw the beginnings of the craft of fine pottery.

From the earliest times the plastic quality of clay has been appreciated by man; it could easily be formed by hand into votive figures or used to seal baskets, and is very widespread in riverbeds and deposits. The history of pottery really starts with man's realization that when clay is fired to a sufficiently high temperature (and an open fire can just achieve this), the nature of the material changes. It can no longer be softened by water to return it to its plastic state. The fragility of fired clay did not lend itself to the nomadic life, however, and the development of pottery began only with the settled agricultural societies of the neolithic period. The earliest known pottery, from Anatolia, dates from the 7th millenium BC.

Early forms of decoration

Except for the fritwares of Egypt and some minor lead-glazed wares in Mesopotamia and the Roman Empire, vitreous glazes were almost unknown in the ancient world, but simple earthenware technology was progressively developed. The clay itself was refined and slips of liquid clay were applied, which, once fired, could be burnished to give a shiny even surface, helping to reduce porosity. It was found that certain mineral colours would withstand the firing process and a limited, sober range of browns, orange-reds and blacks was used. By the 4th millenium BC the potter's wheel was in use in Mesopotamia, which allowed potters to work more quickly and produce greater refinement of form. A better understanding of kiln technology gave greater control to the firing.

The Greeks, as heirs to all these advances, brought earthenware to a very high standard, the limit of what was then possible. An advantage of the absence of glaze is that the sharpness of detail was not obscured. Quite apart from the great work of the Greek potters, it is remarkable how often, in very early times, pots were produced that must be considered by the most exacting modern standards to be fine works of art. Pottery seems to have been the form of the highest artistic expression for many ancient societies. It appears that in the simpler early societies pottery was very highly regarded, whereas in the more advanced societies much of the artistic talent and energy was directed into other crafts such as metalwork and sculpture. The most obvious exception to this is ancient Greece.

One of the earliest groups of fine painted pottery is from Hacilar in Turkey, dating from about 5000 BC; bold geometric designs were painted in red on a whitish slip and were highly burnished. At about the same time in Northern Mesopotamia, Samarra ware was produced using similar techniques – simple human and animal representations were combined with well arranged geometric designs. Also in Mesopotamia the remarkable Tell Halaf wares (4500–4000 BC) reached their peak, with painstaking and elaborate geometric designs in red, white and black on a thinly potted pinkish buff body. These luxurious wares represent a technical high point that was not to be matched in Mesopotamia for nearly three thousand years. In particular, the kiln technology was such that it allowed careful firing; the wares were placed in a chamber, which ensured they did not come into contact with fire and smoke, so retaining their clarity of design.

Further to the east, in Persia, the Susa wares of 4000–3000 BC were notable for the quality of their potting and the judiciously arranged geometric designs on beakers, bowls and chalices. Good painted pottery continued to be made in Persia using the same techniques, but, as in Mesopotamia, the early wares were rarely surpassed. Much later, at Sialk around 1000 BC, new and startling shapes, (such as long spouted ewers) were introduced, often painted with lively designs and animals.

As well as painted wares the ancient world had a long tradition of monochromes whose appeal relied on the strength of their forms and, in the finer wares, on the quality of the burnished surfaces. Amongst the finest of these monochromes were the superb early Anatolian Hittite wares of the 16th–18th centuries BC. The sharply defined forms, inspired by metalware, had striking curved surfaces and abrupt changes of contour, with a highly burnished slip.

Pottery did not play an important part in the arts of

A Panathenaic prize amphora, Athens, c. 560 BC, bearing the inscription "I am one of the prizes from Athens". Amphorae of this type held the oil won by victors at the Panathenaic festivals of 562 or 558 BC; this one was discovered in a tomb in Athens by Thomas Burgon in 1813.

Egypt, but was usually relegated to a functional role. Some of the finest Egyptian wares date from the pre-dynastic period of about 4000 BC; these Badarian wares were thin walled with a well burnished surface, relying for their impact on their simple unadorned forms and lustrous surface. Similar techniques were used with the Meydum wares of 2700–2100 BC to produce some very refined and well potted forms. An important technical step in dynastic Egypt was the discovery of vitreous-glazed quartz fritware; copper was used on this ware to produce an intense turquoise. Fritware lacked the plastic qualities of clay and shapes tended to be simple and thickly potted. The ware was revived in Roman times and was the ancestor of the Islamic fritwares that were to play such an important part in that ceramic tradition.

Minoan and Greek pottery

The Minoan culture of Crete from its beginnings in about 3000 BC to the destruction of the capital Knossos in about 1400 BC, had a lively and varied tradition of decorative arts in which pottery played an important part. The Minoans traded with Egypt, where their pottery was held in high regard. The stylized designs of the Minoan "Middle Bronze Age" developed into a free and very naturalistic style, where plants and marine life were the main focus; no human or historical subjects were depicted. The designs frequently cover the whole surface of the piece, and their influence was very strong in the pottery of Mycenean Greece. The Mycenean potters introduced elements of eastern designs, such as sphinxes, griffins, bulls and birds, but their more disciplined approach lacked some of the freedom of Minoan pottery.

The Greeks used the same basic techniques as the Minoans and Myceneans, but made improvements in the refining of clay, especially the highly burnished black slip. The handsome but restricted repertory of forms was finished with great precision; and later classical revivals of these forms are still widely used to this day. Above all it was the painted figure that came to characterize Greek ceramic art.

The earliest style of Greek pottery, known as geometric, began to emerge around 1000 BC. It reached its peak in Athens around 750 BC with monumental vases which were decorated with bands of elaborate geometric patterns and high stylized figures and animals from both legends and daily life.

The Black Figure style (700–530 BC) was developed first at Corinth but it was again in Athens that it matured into its final form. Border patterns were simplified or even eliminated, allowing a broad band of human figures to dominate the decoration. As the name suggests, the figures were painted in black slip and silhouetted on the orange/buff body; fine detail could be incised through the black to the paler body and some purple, red and white pigment was used sparingly. The subject matter was wide-ranging, including figures from mythology, history and contemporary life, athletics and the theatre. The painter and the potter would often sign their work. The Black Figure style was at its height from 550 to 530 BC.

Around 530 BC the Red Figure style was introduced. This reversed the earlier technique; the background was painted black, with the figures left in the reddish clay. This simple change allowed for greater fluency and finer draughtsmanship; previously detail could only be added by the relatively limited technique of incising the design in lines of unvarying thickness, whereas now the painters could use the brush for detail. The depiction of anatomy and drapery became more subtle and a slightly more light-hearted approach is noticeable. In the earlier wares figures had generally been shown in profile, but it seems that just before 500 BC new ways of illustrating the body were mastered and the figures became more expressive.

The Athenian style was taken to other centres in the Mediterranean, notably the Greek colonies of southern Italy. Artistic and technical standards declined throughout the 4th century BC and by the end of the century Red Figure wares had ceased to be made.

Roman pottery

The Roman tradition of pottery was quite different from that of Greece, and painted decoration is notably absent. The Roman potters had learned the art of producing a very refined clay body and the use of slips from the Greeks, but the wares were fired in an oxidizing atmosphere to produce a rich and bright red surface which could be burnished to a high finish. This "red gloss" pottery was the most widespread within the empire and dominated the production.

The decoration of the finest wares was formed from intricately carved moulds, often based on repoussé metalwork, notable amongst these are 1st century Arretine wares from Arezzo near Florence. Typically, decoration consisted of a sharply defined frieze of figures or hunting scenes. The craftsmanship and the ordered sense of design was excellent. Pieces were often signed, and it is probable that the potters were of Greek origin. The centre of activity later shifted west to Gaul where high technical standards were maintained but artistic quality declined. Potteries followed the expansion of the empire and "red gloss" ware was made at many centres throughout England, the Rhineland and North Africa. Plain wares were plentiful, and also carved decoration, which was made somewhat in the manner of cut glass. Slip-trailed decoration, in the hands of skilled craftsmen, produced refreshingly lively wares, sometimes showing the Celtic traditions of the fringes of the empire.

An Arretine krater, Augustan period, 27 BC–AD 14. This vase, which is missing its foot, is decorated with a design showing the seasons and has the stamp of the potter Cn. Ateius. The finely moulded and burnished red gloss wares dominated the pottery production of the Roman Empire.

Green and yellow lead-glazed wares were made in a number of centres in Asia Minor between the 1st century BC and the 1st century AD. The technique had long been known in Mesopotamia but was not very widespread. The lead glaze was frequently used on moulded wares inspired by silver or gold prototypes. The technique was disseminated through Italy, where it was little used, and spread to Gaul, the Rhineland and even, to a small extent, England. The Byzantine Empire continued to employ lead

A Mochica effigy vessel of a seated woman, Peru, AD 200–500. Many rich and varied ceramic traditions developed in pre-Columbian America, in complete isolation from the rest of the world.

glazes and eventually reintroduced their use to Europe in medieval times.

The Americas

The only major civilizations to have developed in complete isolation from the Western and Asiatic civilizations are those of pre-Columbian America, where the Bronze Age continued until the Spanish conquests of the 16th century. Many cultures came and went throughout the ages and due to the widespread custom of burying pottery with the dead, they have left a rich ceramic legacy. Glazes and the potter's wheel were never invented and so all pottery is hand-formed.

In the Andean region of Peru an abundant and varied body of work was produced. The early Chavin culture created finely structured wares with noble proportions; this was followed by the Mochica (*c.* AD 1–1200). The Mochica produced some of the most sensitive modelled portraiture of any culture, in the form of stirrup cups which display a startling realism. The Nazca and later Inca wares were notable for their brightly coloured decoration, an exuberant and now mysterious iconography having been executed with a high degree of draughtsmanship.

Mexico and Central America had a bewildering array of ceramic traditions. Finely burnished monochromes and ornately decorated wares with grotesque and terrifying motifs were produced to a very high standard. Zapotec and Mayan sculptural pottery had a powerful if ornate beauty, whilst the early Olmec models are more naturalistic and gentle in character.

The less sophisticated cultures of the Pueblo Indians of North America produced attractive painted wares with well proportioned geometric and animal designs.

INFLUENCES

1

*A section of the west frieze of the Parthenon (**1**), 477–432 BC, with horsemen and attendants preparing to form the Panathenaic procession. The great developments in sculpture and painting, of which little survives, were closely followed by the painters of the Attic red and black wares.*

2

*Roman glass from Syria (**2**) 64 BC–AD 31. Glass fulfilled many of the same functions and required much the same technology as pottery. A pottery glaze, which was in effect a glass-like covering of the earthenware body, was used sparingly in the Ancient World.*

3

*A silver cup (**3**) with repoussé design showing scenes from the history of Troy; Roman 1st century AD. Roman "red-gloss" and lead-glazed wares followed the form and decoration of silver prototypes very closely.*

4

*A glass bottle (**4**), Syria, 1st century AD. Glass workers had refined both decoration, and technique. Pottery also required an understanding of mineral colours and a means of producing high temperatures.*

5

*A fresco (**5**) from the Hall of Mysteries in Pompeii, c. AD 79. Roman potters derived much of their decorative inspiration from the painters and sculptors who were at the forefront of artistic developments.*

EARLY CIVILIZATIONS

*Tell Halaf ware (**1**) and (**2**) from northern Mesopotamia, 4500–4000 BC. Carefully contrived geometric patterns were used in a range of colours on a pinkish-buff body. These elaborate early wares were not equalled in Mesopotamia for nearly 3000 years.*

Oinichoe (***3***) *from Aegina, Cycladic Islands, early 7th century BC. The animal frieze and bead show strong Eastern influence.*

1

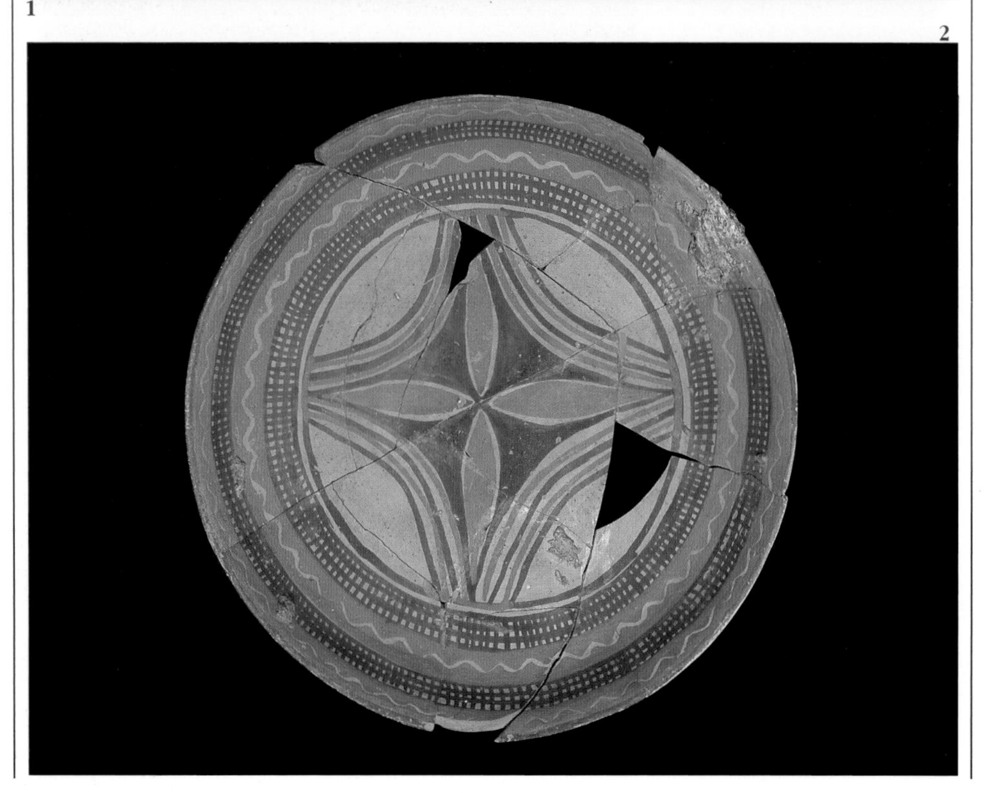

2

3

*A Susa ware beaker (**4**) from Persia, 4000–3000 BC. Amongst the most distinguished of the early wares, Susa pottery combined an extremely fine fabric with carefully executed geometric designs, allied to a range of strong simple forms.*

4

5

*A Mycenean goblet (**5**), c. 1300 BC. The octopus was also a popular subject on Minoan painted pottery, which influenced the pottery of the Greek mainland.*

A Mycenean krater *(**6**), 1300–1200 BC, used for mixing water and wine. Animals were often subjects on Mycenean pottery.*

6

GREECE

1

2

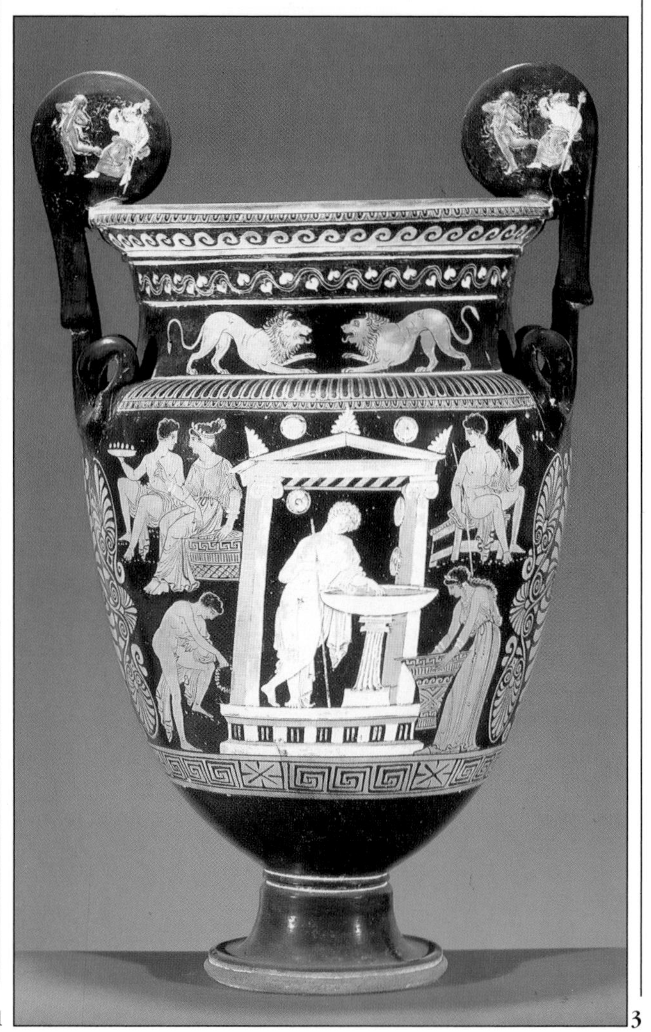

3

4

5

6

*A late geometric pitcher (**3**) showing mourners at the laying-out of the dead, Athens c. 720-700 BC.*

*A Red Figure plate painted by Epiktetos (**2**), Athens, c. 520–510 BC, showing two revellers. A neck-amphora (**6**) in the Red Figure style, attributed to the Berlin painter, Athens c. 490 BC.*

*The Greek potters of southern Italy initially followed the styles of Athens, as in this krater (**3**), Apulia, c. 380–370 BC, attributed to the Iluipersis painter.*

*A black figure neck-amphora (**4**) by Exekias, Athens 540–530 BC.*

*An elaborate Canosan askos (**5**) with moulded figures of Victories, Apulia 3rd century BC. The Red Figure technique had finally disappeared by the end of the 4th century BC.*

ROME

1

*Red gloss ware dominated Roman pottery. The finest were the Arretine wares (**2**) such as this bell-shaped* krater, *c. 10 BC, which were made at Arezzo in Italy and used moulded decoration of Hellenistic origin. Later the centre of production moved to southern Gaul (**1**) where fine quality, if less well designed, moulded wares were made.*

3

2

4

*The brilliant turquoise glaze on Egypto-Roman fritware (**3**) was derived from copper.*

*Green or yellow lead glazed ware (**4**), c. 1st century BC–1st century AD. This was made initially in Asia Minor, and later throughout both the Roman and Byzantine empires.*

*Red gloss cup (**5**), Gaul, 2nd century AD. This type of incised decoration imitates cut glass.*

5

PRE-COLUMBIAN POTTERY

1

*A Mochica stirrup spout portrait jar (**1**), Peru, AD 200–500. The Mochica culture produced some of the most startlingly realistic portraiture found in pottery. Although made from moulds, the pots were further decorated to show individual characteristics.*

*A painted flask (**2**), Aasazi culture, Pueblo period, c. AD 1000. The early pottery of North America excelled in angular painted decoration.*

2

3

*Two Aztec painted orange ware bowls (**3**), Mexico, 14th–15th century AD.*

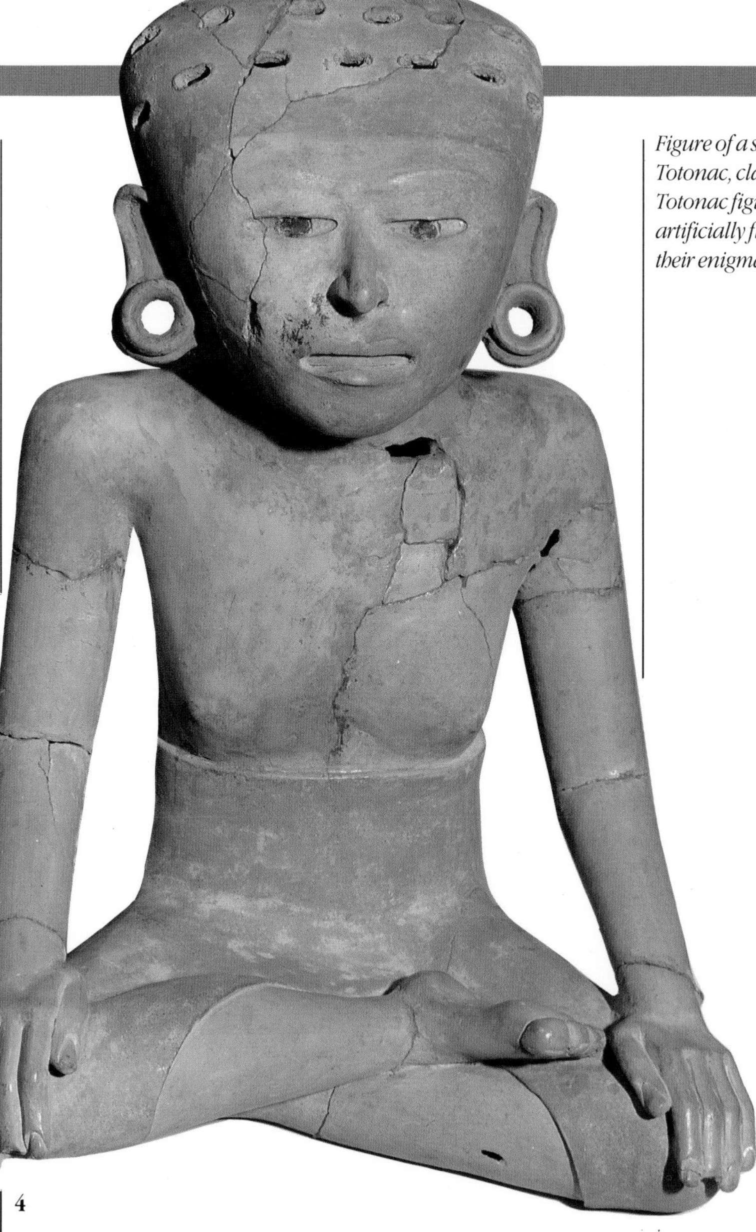
4

*Figure of a seated woman (**4**), Totonac, classic period, Mexico. Totonac figures are noted for their artificially flattened heads and their enigmatic expressions.*

*A stirrup vessel (**6**), Chimu, Peru, painted with the dragon god. The stirrup-shaped handle formed as a spout persisted for a period of over 2000 years on the north coast of Peru.*

*A painted bowl (**7**), Maya, classic period, Mexico. Narrative paintings illustrating epic stories of the underworld were often shown on grave-goods.*

6

7

8

5

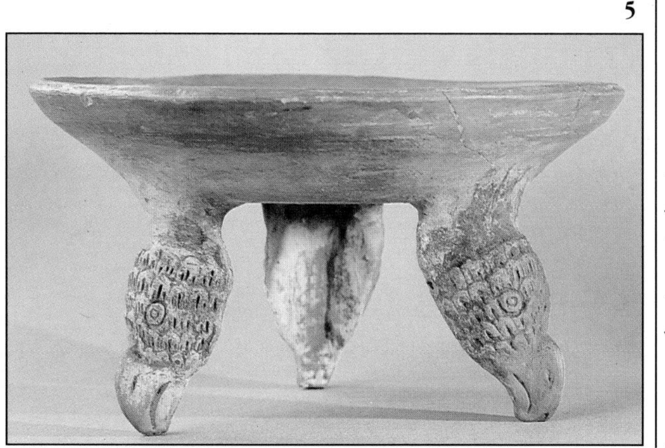

*A painted bowl with feet in the form of eagle heads (**5**), Aztec, Mexico.*

*A bowl painted with human figures (**8**), Mogollon, Mimbres valley, North America, c. AD 950. The bowl has been deliberately broken to release the owner's soul at his death.*

A Guan yao vase, southern Song dynasty, 12th–13th century.

CHAPTER · TWO

CHINA

"Like bright moons cunningly carved and dyed with Spring Water. Like Curling discs of thinnest ice, filled with green clouds."

HSU YIN, TANG DYNASTY ON YUE WARE

CHINA

No more noble a tradition of ceramics exists than that of China. It was here that the first high-fired stonewares, which are resonant and non-porous, were produced, and the earliest porcelains predate those of any other culture by little short of a thousand years. As remarkable as the technical achievement is the artistic skill which produced, after the vigour and splendour of the Tang dynasty pottery, the exquisite refinement of the Song dynasty, whose celadons and other monochromes are, even today, the yardstick by which many contemporary artist potters judge their work.

A partially glazed figure of a court lady, Tang dynasty, first half of the 8th century AD. The funerary sculpture of the Tang dynasty brings the court of the period vividly to life. The glaze here includes the uncommon early use of cobalt blue that was to dominate later ceramic production.

Early stoneware

The earliest Chinese pottery of note dates from the Yangshao culture during the Neolithic period. It combined strong forms and vigorous geometric designs. Accomplished pottery was also produced by other cultures in the vast and, as yet, ununified country. The Shang dynasty of the 18th to the 10th century BC excelled in the production of quite remarkable bronze vessels that are the technical and artistic equal of anything in antiquity. Bronze casting required very high temperatures and the use of moulds – such techniques must have contributed to the development of stoneware. Some good pottery was produced during the Shang dynasty, though its occurrence is rare; its forms and decoration are closely related to the bronzes of the period.

From the period of the late Warring states and early Han dynasty in the 3rd and 4th centuries BC high-fired stonewares were produced that began to take on truly ceramic forms, though sometimes details such as *taotie* mask handles (stylized monsters) still show a debt to contemporary bronze casting. Many of these early stonewares have feldspathic glazes of an olive-green to brown colour, and these can be considered the ancestors of the celadon glazes which formed such a crucial element in the Chinese ceramic tradition.

Most of the early surviving ceramics have been excavated or robbed from tombs. The Chinese tradition of lavishly endowing the tombs of notables and royalty with earthenware figures and vessels was dramatically illustrated in 1974 by the chance discovery of the vast terracotta army of the Emperor Qin Shihuangdi, who unified China in 221 BC. The breathtaking scale of the terracotta army and the artistic accomplishment of the individually modelled soldiers and horses astonished the world, and it has as yet only been partially excavated. The Han dynasty has long been known for its wide variety of funerary wares, many of which can be seen in museums. These comprise human and animal figures, as well as domestic and ritual vessels. Some are of painted earthenware and many others have good brown or green lead glazes which can frequently be seen in excellent states of preservation. It can be assumed that similar vessels were in domestic use.

The following period, known as the Six Dynasties (AD 219–581) was a period of some political confusion in Chinese history. Nonetheless important developments in ceramics occurred; the Yueh stonewares continued the earlier tradition of feldspathic glazes and this ware reached new artistic heights. The refinement achieved in these high-fired wares was to lead to the development of porcelain in the succeeding Tang dynasty.

The development of porcelain

The Tang dynasty (AD 618–906) ushered in an era of stability and prosperity conducive to artistic development. Trading and cultural links were expanded to the Middle East and Japan. It was a golden age for poetry and painting. Pre-eminent amongst the ceramic works of the Tang dynasty are wonderfully modelled and glazed horses, camels and a wide diversity of human figures – the bearded Middle Eastern traders, court ladies, dwarves and musicians suggest some of the variety of this confident and cosmopolitan society. The Tang potters could also create masterpieces of simplicity, as can be seen in the finely proportioned jars, glazed in single colours or splashed with cleverly controlled *sancai* (three-colour glazes).

Finely potted white porcellaneous wares were developed in the 7th and 8th centuries; these were simple unadorned forms, usually dishes or vessels, often based on silver shapes. It is worth noting that the Chinese do not make the clear distinction between fine stonewares and porcelain that is made in the West.

The Song dynasty, AD 960–1279, is regarded as the classical age of China. The Song dynasty brought an unrivalled aesthetic refinement to ceramics; it was a period which produced some of the greatest masterpieces of the potter's art. Decoration was reduced to a minimum, and forms of perfect harmony were created, with unrivalled monochrome glazes.

Little earthenware of note was made in the Song dynasty; the emphasis was on the higher fired stonewares and the further refinement of porcelain. In the early part of the dynasty the most important centres of ceramic production were in the north. Notable amongst these wares are the ivory-white Ding wares and the "Northern Celadons", both of which are finely potted and decorated with floral sprays and meanders of great beauty freely carved into the body. Moulded decoration was also common. The Jun wares have no incised decoration, relying entirely for their effect on the strength of their forms and the opaque lavender-blue glaze with occasional crimson splashes. The rarest and finest of Song ceramics is Ru ware, made for the Northern court. The forms are either simple bowls and dishes or more complex shapes derived from bronze and lacquer ware. They are covered in a subtle greyish blue glaze which is frequently crazed.

In 1127 the Jurchen Tartars invaded the Northern provinces and forced the Song court to retreat to the south, where the dynasty continued until 1279. Ceramics continued to flourish here, particularly at Longquan, where the difficult process of firing celadons was perfected. The best of these glazes exhibit an almost sky-blue tone. For the Court, Guan (official) wares were produced, which are frequently articles for the scholar's table such as brush washers under intentional and carefully controlled crackled glazes. Yingqing porcelain, with its pale blue glaze and thinly potted body, was widely made and equally widely exported. The city of Jingdezhen began its long domination of Chinese porcelain, which has lasted to the present day.

This imperial yellow-ground landscape bowl, Qianlong mark and period, 1736–1795, is a particularly brilliant example of the imperial taste of the period.

A Ming blue and white dish, Chenghua mark and period, 1465–1487. The formal wares of this period achieved a high level of refinement which later potters strove to emulate.

Blue and white porcelain

The Mongol conquest of China under Khublai Khan, the founder of the Yuan dynasty (1260–1368), brought about a new artistic climate. The extreme refinement of the Song gave way to a more vigorous, bold and energetic style. The great innovation of the Yuan period was the underglaze decoration of porcelain, where colour is applied to the clay body and integrates with a transparent covering glaze. Few colours could withstand the high firing temperature of porcelain – by far the most successful was cobalt blue, whose use as an underglaze colour was perfected in a short time. Cobalt blue was used for a short while on Tang pottery at least as early as 723, but it was possibly reintroduced to China from Persia in the 13th and 14th centuries. Blue and white porcelain now became the most famous and influential of all Chinese ceramics. One other underglaze colour, copper-red, was attempted for a short period but it proved very difficult to control and was abandoned, to be revived only intermittently in later centuries. At its best the results achieved with copper red were spectacular, and these wares are now amongst the most sought after of all porcelains.

Yuan blue and white is densely decorated and powerfully painted with bands of floral meanders, dragons, phoenixes and other creatures. The dishes and rarer vases are often massive and were exported widely to South-East Asia, India and the Middle East.

With the foundation of the Ming dynasty in 1368, a Chinese dynasty was once again in power. The blue and white porcelains of the 15th century constitute another great classical age in Chinese ceramics; these wares achieved heights of technical and artistic excellence that have never since been equalled. The beautiful designs of the Yongle period (1403–1424) gave way to the more formal Xuande wares, and in the reign of the Chenghua emperor (1465–1487) pieces of great delicacy were made, notably those known as the Palace bowls.

Monochrome glazed wares of Imperial yellow, copper-red and blue were made, sometimes with scarcely visible incised decoration. In the Chenghua period the first and rarest of the enamelled wares (where decoration is applied over the glaze) were made in a style known as *doucai* or "clashing colours".

The later Ming period, until its final collapse in 1644, showed a decline in technical standards, although very lively blue and white and enamelled pieces continued to be made in great profusion and variety. One especially successful group of wares is known as *wucai* or "five colour", and these have been particularly popular and influential in Japan.

Other notable centres of production in the late Ming period were at Swatow, where blue and white and polychrome dishes were made which at their best are very striking, and in Fujian province where the production of blanc-de-chine, a fine ivory or bluish-white porcelain, was started. The finest blanc-de-chine pieces were made at Dehua and the tradition has continued to this day; most notable amongst the productions of these kilns are the beautifully modelled figures of Guanyin, the "goddess" of mercy, and other religious figures.

With the fall of the Ming dynasty a new era of experimentation began, known as the Transitional period. Blue and white porcelain continued to dominate the production. New shapes were developed and the standard of potting and painting is frequently of a very high order. Some of the production was for the domestic market but increasing quantities were exported to the West along with the late Ming style porcelain that was now in decline.

Production for export

With the decline of Imperial patronage, new markets were developed – Japan and Europe became important centres for export. The Portuguese were the first Europeans to trade with China in the 16th century but in the early 17th century the Dutch and then the English began to take over. Vast quantities of pots were shipped to Europe, where the fashion for blue and white reached a peak in the late 17th century. Most of the shapes were designed specifically for the European market.

After the ravages of civil war and the establishment of the Qing dynasty (1644–1912) porcelain production at Jingdezhen was reorganized. The last great phase of

A late Ming kinrande *cup, mounted in silver-gilt for the von Manderschneidt family, c. 1583. This cup, brought from Turkey, was one of the earliest porcelains to appear in Europe. The pieces were highly treasured, as no material comparable to porcelain was then known.*

Chinese ceramics now began, which reached a peak in the first half of the 18th century. In the later 18th century and throughout the 19th, the quality of potting, painting and the porcelain itself, declined significantly. Most of the ceramics of this period can be divided into two groups, the pieces for export and those for the domestic market.

The West increasingly demanded coloured wares. The range of enamels known as "famille verte", with its brilliant translucent colours, dominated the production of domestic tableware until the 1720s when a pink enamel was introduced from the West. This became the basis for a new palette known as "famille rose". A vast range and quantity of wares were produced: tea and dinner services, vases and special commissions. Armorial dinner services were ordered through the East India companies; designs were frequently copied by the Chinese from Western prints. Much magnificent porcelain was produced but rarely of the same quality as that reserved for the domestic market.

The Imperial kilns continued many of the classic Ming traditions of blue and white and produced an expanded range of monochromes. During the Yongzheng period (1723–1735) technical standards almost rivalled those of the 15th century. With the arrival of the famille rose palette, exquisite Imperial wares were developed which must rank as the masterpieces of the period. But towards the end of the Qianlong period (1736–1795) the standards declined – there was a tendency to overdecorate wares and a general lack of invention.

Within the general orbit of the Chinese civilization there were other centres of production, such as the Kingdom of Annam (now Vietnam), where during the 15th century handsome blue and white wares were made in imitation of the Chinese. Korea has a long tradition of ceramics and in the 12th and 13th centuries produced beautiful celadons that rival those of the Song dynasty. The Koreans were the first to use the technique of inlaid decoration, and the later ceramics of Korea had decisive influence on those of Japan.

An underglaze copper-red dish, Yuan dynasty, 14th century. The elusive red derived from copper was extremely hard to control and was soon abandoned in favour of blue and white designs. Copper-reds are amongst the rarest of Chinese porcelains.

INFLUENCES

*An archaic bronze wine vessel or gu (**1**), Shang dynasty 12th–11th century BC. The remarkable flowering of bronze casting in ancient China developed some of the skills necessary for firing stonewares and the bronze shapes were repeatedly copied in ceramics.*

*A section of wallpaper (**2**), 18th century, showing the manufacture of porcelain and a merchant examining the wares.*

*A carved cinnabar lacquer dish (**3**), 14th century. The superb sense of design shown in the finest porcelain was also matched in the other art forms.*

1

2

3

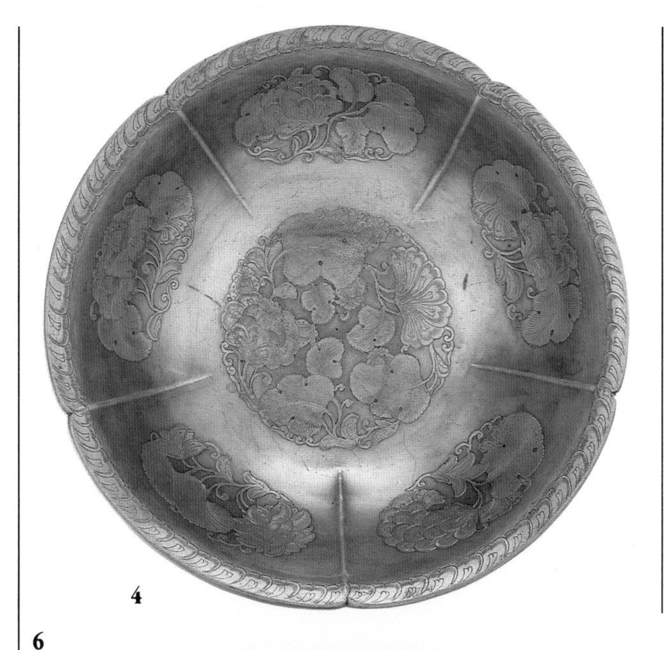

4

*An engraved and parcel-gilt silver bowl (**4**), Tang dynasty, 8th century AD. This shape is found in the porcelains of the succeeding Song dynasty.*

*A jade peach-shaped brush washer (**5**), early 18th century. The almost mystical reverence in which jade was held inspired the yueyao and celadon potters to emulate it, in the quality of their glazes.*

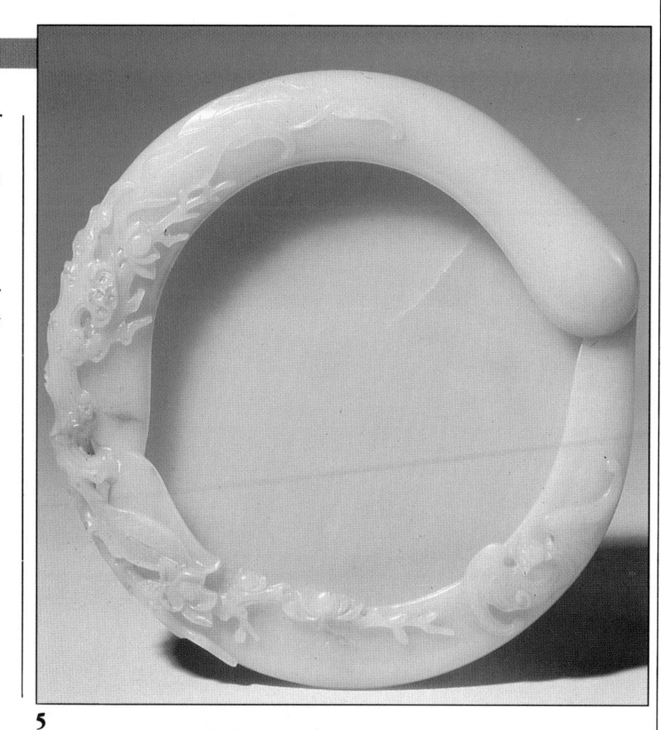

5

6

*A detail of a hanging scroll (**6**) attributed to the 12th century Emperor Huizong, showing scholars meeting in a garden to enjoy food and conversation. The extraordinary refinement of the Song court inspired some of the most sublime pottery ever created; simplicity was combined with an exquisite sense of form, whilst decoration was often minimal.*

EARLY WARES

*The large jars of the Yang-shao culture from Kansu province (**1**) of the 3rd millenium BC with their bold curvilinear decoration are amongst the most striking of all neolithic pottery. The black and red decoration on the well levigated body was often highly burnished. The designs were usually geometric, but occasionally stylized human and animal forms were used.*

1

*A green-glazed red pottery watchtower (**2**) of the Han Dynasty, (206 BC–AD 220). Multi-storeyed towers were built inside the courtyards of Chinese farmsteads; realistic models of these and other farm buildings, such as pig sties, are often found in Han Dynasty graves.*

2

3

*A stoneware jar (**3**) of the Warring States period, 481–221 BC. These jars with regularly spaced impressed designs can sometimes be found with primitive glazes.*

4

*A glazed red pottery figure of a seated dog (**4**) of the Han Dynasty (206 BC–AD 220). The Han sculptural funerary pottery was often of a very high order and foreshadows what was to come in the Tang Dynasty.*

TERRACOTTA SOLDIERS

The vast terracotta army discovered in 1974 in the mausoleum of the first Emperor of unified China, Qin Shihuangdi, who died in 210 BC, is possibly the most extraordinary archaeological find of the 20th century. Comprising over 7000 lifesize soldiers, horses and chariots individually sculpted with remarkable realism and sensitivity, the awe-inspiring scale and quality of the work has no parallel inside or outside China.

TANG

*A green glazed tripod jar and cover (**1**). The fine glaze and precision of the potting are typical of the best Tang pottery. The shape derives from metalwork and even details such as the rivets on the cover are copied.*

*A sancai (three-colour) glazed figure of a Bactrian camel (**2**). The vigorous and realistic modelling is typical of this dynamic period in Chinese art.*

*A finely modelled figure of a seated lady (**3**), holding a lion cub with a funnel in its mouth.*

1

2

3

A green-glazed rhyton *(**4**) or drinking vessel. The beaded ground suggests it was copied from the ring-matting of contemporary silver.*

4

5

6

*The carefully controlled use of splashed glazes (**6**) on the well-potted Tang forms produced some of the most satisfying Chinese pottery.*

A fine sancai *figure of a caparisoned Fereghan horse (**5**). More than any other animal the horse enjoyed the attention of the Tang potters.*

CELADON

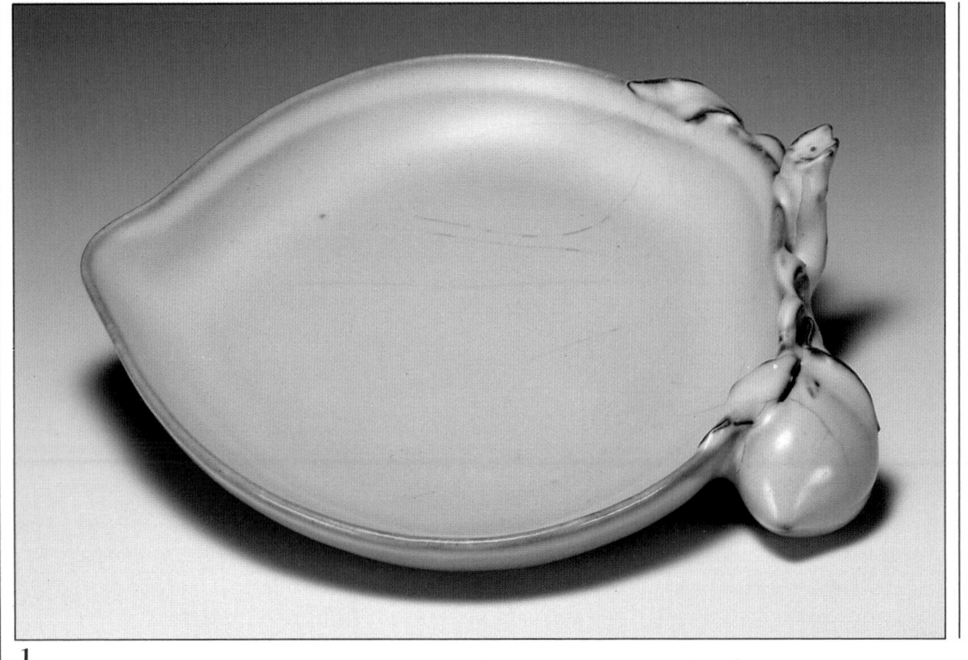

1

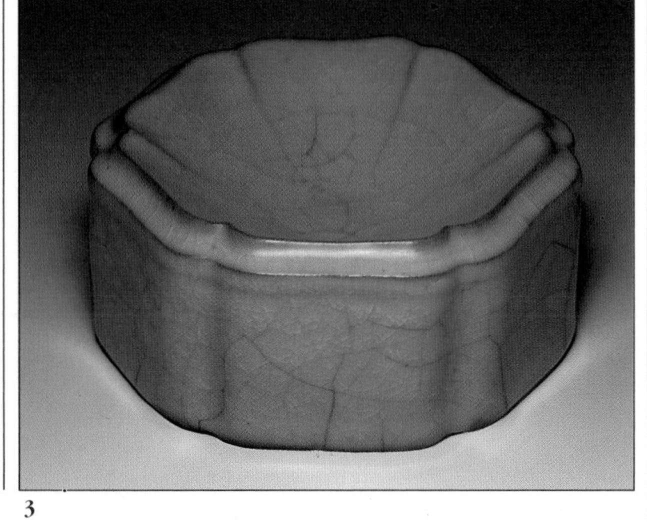

3

2

*Amongst the most refined of all Chinese ceramics are the official wares made for the courts of the Song dynasty emperors. The fine crackle and jade like colour of this late 12th/early 13th century Guanyao brushwasher (**3**) are characteristic of these rare wares. Guan type glazes, as shown on this peach-shaped brushwasher (**1**) of the Yongzheng period (early 18th century) were periodically revived in imitation of the earlier wares.*

*A Northern Song celadon box and cover (**2**) of the 11th or 12th century, with characteristic olive glaze and superb fluent carving.*

4

5

6

7

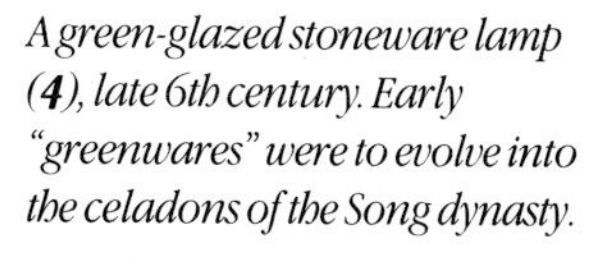

*A green-glazed stoneware lamp (**4**), late 6th century. Early "greenwares" were to evolve into the celadons of the Song dynasty.*

*A large carved celadon dish (**5**), early Ming dynasty. Dishes of this type were exported to the Middle East throughout the Yuan and Ming dynasties.*

*A Yueyao bowl (**6**), 10th century. Yueh wares were continuously refined, reaching their peak towards the end of the Tang dynasty.*

*A Ru ware bowl (**7**), early 12th century. The rarest and finest of all the official wares, Ru ware was briefly made for the Northern Song court.*

SONG

1

2

3

*A Jian yao "hare's fur" teabowl (**1**), 12th–13th century. These teabowls were made in South China for local domestic use, but were imported to Japan by Zen Buddhist monks. They became much admired and were used in the tea ceremony; they are known in Japan as tenmoku wares.*

*A Henan jar (**2**) of the 12th–13th century from north China. These handsomely potted wares were frequently painted in a loose and spontaneous calligraphic style and had a considerable influence on the early British studio potters such as Bernard Leach.*

*A Guan or "official" ware dish (**3**), 12th–13th century. These remarkably restrained wares were popular in the highest courtly circles of the Song Dynasty.*

*A Yingqing vase (**4**), 12th century, an early and beautifully potted example of the ware that was the precursor to the blue and white porcelains of the Yuan Dynasty.*

4

*The Ding yao kilns of north China perfected an ivory-white porcelain (**6**) in the 11th and 12th centuries, with fluently incised decoration.*

*A Cizhou jar (**5**), 12th–13th century. A range of techniques were used with the country stonewares of north China to achieve works of great artistry.*

*A Jun yao dish (**7**), 13th century; carved decoration was not used on Jun wares, which were either left entirely plain or decorated with irregular crimson splashes.*

5

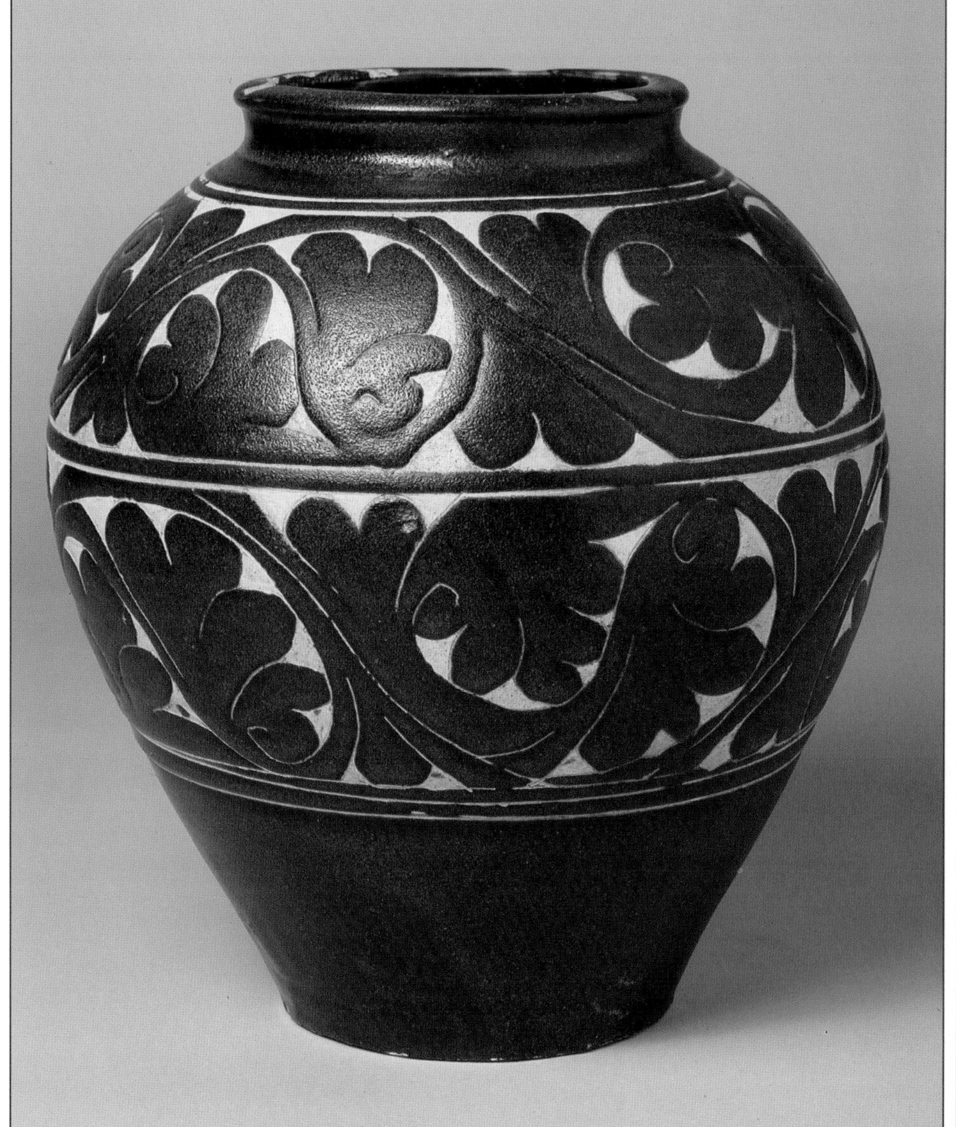

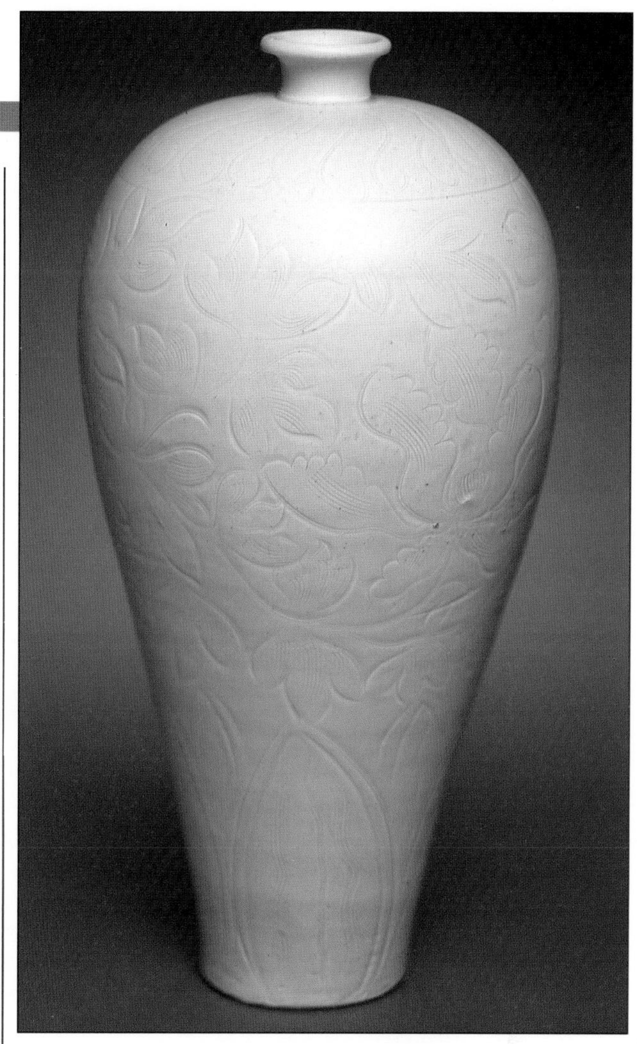

6

7

YUAN

1

*A large blue and white dish (**1**), mid 14th century. In a remarkably short period the use of cobalt blue on porcelain developed into the mature Yuan style, as exemplified in this dish. The vigour and boldness of the styles associated with the Mongol invaders contrasts with the elegant refinement of the Song dynasty Chinese whom they supplanted. The scale of their works was typically much larger than the Song wares which were often of a size that could be easily handled.*

*An underglaze copper-red ewer (**2**), Yuan or early Ming dynasty. The production of blue and white porcelain was easily mastered by the Yuan potters, whilst the elusive copper-red required a carefully controlled reduction firing. This entailed restricting the flow of oxygen into the kiln and the final result left much to chance. The muted tone of the red on this ewer, which is characteristic of much of this type of ware, indicates a slightly incomplete reduction of the copper.*

2

*A large wine jar, (**3**), Guan, c. 1330–1340 AD. This remarkable jar, which is one of only four known, exhibits a combination of decorative techniques found on no other types of ware. The use of cobalt blue and copper red are combined with relief decoration of a kind that is sometimes found on late Yingqing porcelains.*

*A Longquan celadon dish (**4**), 14th century, decorated with an underglazed dragon relief. The unglazed decoration is a variation on the normally completely glazed celadons that found particular favour in the Middle East.*

3

4

*A large blue and white jar, (**5**), Guan, mid 14th century. The vigour of the Yuan designs, characterized by the spiky and dynamic floral meanders, was softened and formalized under the succeeding Ming dynasty.*

5

1

2

*An early Ming blue and white dish (**1**) and (**2**), Yongle period. The quality of execution and draughtsmanship allied with classic formal designs make the early Ming blue and white porcelains the finest ever produced. The splendid vigour of Yuan painting developed into the more controlled approach of the 15th century. Dishes of this type were much admired by rulers in the Islamic world and fine collections have survived in the Topkapi Serail in Istanbul and at the Ardebil Shrine in Iran. Potters in the 18th century tried to recreate these wares but in spite of their considerable technical skill the majestic quality of these early wares eluded them.*

A jar painted in doucai *(clashing colours) enamels (**3**), Chenghua mark and period, late 15th century. These are the earliest enamelled Ming porcelains.*

3

4

*A palace bowl (**4**), Chenghua mark and period. The Chenghua porcelains achieved an unrivalled refinement.*

An early Ming meiping *(**5**), Yongle period, early 15th century. The cobalt blue shows the "heaping and piling" effect of early Ming blue and white that later copyists tried unsuccessfully to emulate.*

5

16TH AND 17TH CENTURIES

*A famille verte vase (**1**), Kangxi period, 1662–1722. The brilliant famille verte palette was developed towards the end of the 17th century and much ware was exported to the West by the East India companies. The narrative style of painting often depicts scenes from history and literature.*

1

2

*A beaker vase (**2**), Transitional period c. 1640. At the end of the Ming dynasty the collapse of imperial control of the kilns ushered in a period of renewed experiment and freedom of expression in the painting of ceramic wares.*

*A Swatow dish (**4**), late 16th century. The provincial kilns north of Swatow made boldly painted blue and white and polychrome wares mainly for export to Japan and South East Asia. This dish depicts European ships and a compass.*

3

*A blanc-de-chine figure of Guanyin (**3**), 17th century. The kilns at Dehua in Fujian province avoided painted decoration, preferring to emphasize the beautiful quality of the porcelain itself. Both functional and ritual wares were made but the greatest achievements were in the sculptural religious figures. Amongst these the most popular were the figures of Guanyin, sometimes described as the Goddess of Mercy.*

4

5

A wucai *ewer (**5**), Wanli period, 1573–1619, and a double gourd* kinrande *vase (**6**), 16th century. "Wucai" means five colour decoration; a related type of decoration, characterized by gold in iron-red was known in Japan as "kinrande". These types were amongst the bolder decorative porcelains which became popular in the late Ming period. They found particular favour in Japan, where many fine examples have been preserved.*

6

QING IMPERIAL

*A flask painted in famille rose enamels (**1**), Yongzheng mark and period, 1723–1735. The early 18th century marked the high point of a revival in the production of Chinese porcelain, when the finest wares came close in quality to those of the 15th century. The famille rose palette, whose pink had been introduced from Europe, dominated polychrome decoration.*

1

2

*A pink ground famille rose teapot (**2**), Qianlong mark and period, 1736–1795. In the second half of the 18th century courtly taste favoured elaborate decoration; here there is a somewhat incongruous mixture of motifs, such as the Buddhist lotus and the archaistic dragons derived from early bronzes.*

3

*A peachbloom-glazed chrysanthemum vase (**3**), Kangxi mark and period, 1662–1722. Scholarly taste is reflected in the simplicity of this vase for the writing table.*

4

A famille rose vase (**4**), Qianlong mark and period, 1736–1795. Just as Europeans were captivated by the idea of the exotic East, the Chinese also occasionally used decoration from European sources. This vase was brought from China over a century ago; it was then lost at sea off Scotland and later salvaged from the wreck.

5

6

*Two views of a famille rose cup (**5**) and (**6**), Yongzheng mark and period, 1723–1735. It is painted with lines from a poem reading "At the sighing of the wind in the reeds, the lone flying (goose) calls out", and is typical of the most refined imperial taste.*

EXPORT WARES

1

2

*A pair of famille verte figures of a European lady and gentleman (**1**), c. 1700–15. These figures are traditionally said to represent Louis XIV and Madame de Maintenon and their clothing conforms with French courtly fashion of the period. The Chinese manufacturers eagerly undertook commissions from the European merchants, and these were often executed in a delightfully refreshing naïve style.*

*A Ming blue and white pilgrim flask (**2**), c. 1580–95, painted with the Spanish Royal arms. The arms are those of Philip II and the flask was probably ordered by the Portuguese merchants after the union of Spain and Portugal in 1580. The Portuguese were the first Europeans to trade directly with China by sea.*

*A famille verte dish (**3**), c. 1710, painted in the centre with the arms of Zeelandt. A series of dishes was ordered by the Dutch East India Company with the arms of the provinces and cities of the Netherlands. The Dutch dominated much of the East India trade in the 17th and early 18th centuries.*

3

*A famille rose and grisaille decorated punch bowl (**4**), c. 1802, painted with the Ironmongers Company Hall on one side, the other with the Lord Mayor's coach before the Mansion House. The designs are taken from 18th century prints of London which have been copied very faithfully. Wealthy patrons could order designs of their own choice from the East India Companies; the most common special commissions were for armorial services.*

4

KOREA AND SOUTH EAST ASIA

1

*Annamese polychrome dish (**1**), 16th century. In the 15th and 16th centuries the potters of Annam (now Vietnam) produced stonewares that were close copies of the slightly earlier Chinese porcelains. This 16th century double-headed duck* kendi *(**4**), is an example.*

*A Korean porcelain jar (**2**), 17th-18th century, decorated in underglaze copper-red and showing the Korean potters' mastery of this difficult colour.*

*The Korean celadons of the 12th and 13th centuries (**3**), were strongly influenced by Chinese Song celadons, but new techniques such as inlaid decoration were developed.*

2

3

4

Nabeshima dish, early 18th century.

CHAPTER · THREE
JAPAN

"Its accoutrements and observances are exquisite embodiments of what is most essential in the literature and fine arts of the Japanese people."

KIKUSABURO FUKUI
ON THE TEA CEREMONY

JAPAN

Of all the world's great ceramic traditions, that of Japan is perhaps the most aesthetically challenging and mysterious to a Western observer. In other parts of the world potters generally strove to achieve a level of technical perfection which would raise their works as far above crude "rustic" pottery as possible. The aesthetic intent of the most revered Japanese pottery was often to create a rugged and noble simplicity devoid of artificial refinements; this attitude is evident in the earliest stonewares and became formalized in the tea ceremony wares of the 16th century. A Westerner confronted for the first time by a raku teabowl, with its simply potted, irregular form and uneven thick glaze, would well be forgiven for dismissing it. He would be surprised to learn that it might have been the work of a master potter, given its own poetic name and, as the treasured heirloom of a noble family, used on only the most solemn occasions. The tea ceremony and its masters were central to the development of Japanese taste in ceramics – and eventually a range of wares was accepted as appropriate for the ceremony, including many that were brightly and boldly decorated.

There were also important groups of ceramics apart from the tea ceremony wares – the finest porcelains, such as the rare early Nabeshima wares, achieved a perfection comparable to the Imperial wares of China. Perhaps the best loved wares in the West are the Kakiemon porcelains, which combined one of the most successful decorative palettes ever devised with a beautiful milky-white body.

In the late 17th and early 18th centuries European royal and noble collectors sought these Kakiemon wares avidly. The pieces acquired by Queen Mary II, the consort of William III are still at Hampton Court Palace. The Elector Frederick Augustus I of Saxony, who was to found the Meissen porcelain factory, had a vast collection assembled in Dresden. In 18th century Europe Japanese porcelain was held in higher regard than that of China, and Kakiemon wares were one of the most important influences on the early European porcelain factories. Copies of Chinese wares, however, are comparatively scarce.

Japan had two neighbours – China and Korea – with very strong ceramic traditions, so it is not surprising that techniques were imported and that the other two countries also exerted a strong artistic influence on Japan. Indeed, immigrant Korean potters were responsible for major innovations. It is perhaps more remarkable that Japan developed such a strikingly independent tradition of its own.

Early pottery

The early history of Japan is not well understood, but fine Neolithic pottery was made, notably in the Middle Jomon period of about the 2nd millenium BC. These striking and somewhat bizarre pots have strong asymmetric forms.

A Kakiemon jar and cover, late 17th century. The most beautiful of the porcelains made for the Western market were the Kakiemon wares. A number of Arita kilns also perfected the milky-white "negoshide" porcelain and the characteristic enamels in the late 17th century. Traditionally Sakaida Kakiemon is said to have learned the skills of enamelling from a Chinese in Nagasaki in the early 17th century, but the palette and the designs which developed are unmistakably Japanese. Kakiemon wares were avidly collected by the wealthiest Europeans and were to have a greater influence on early European porcelain than any other Oriental porcelain.

The large *haniwa* pottery figures, of the 5th and 6th centuries AD, depict warriors in full armour, horses and also arresting sculptures associated with the burial tumuli of the period.

An important introduction from Korea in the 5th and 6th centuries AD were the high-fired Sue stonewares – these are hardly distinguishable from the contemporary Silla wares of Korea. The hard stoneware bodies often show irregular splashes of a primitive green-brown celadon glaze, caused by wood-ash falling on them in the reducing atmosphere of the kiln. These chance effects were appreciated and frequently used in later periods, such as on the Shigaraki jars. Contact with China in the Nara period (AD 710–784) brought about increased cultural influences, and close copies of the Tang *sancai* lead-glaze wares were made. With the severing of contact with China the craft of pottery declined somewhat and it was only in the 13th century that a revival took place. This was, in a sense, the start of the Japanese pottery tradition.

During the Kamakura period (1185–1333) there were six centres of pottery known as the "six old kilns": Seto, Tokoname, Shigaraki, Tamba, Bizen and Echizen. Legend states that Toshiro, known as "the father of pottery", settled at Seto after a visit to China and, finding suitable clays, started a pottery. The early Seto wares are influenced by Chinese late Song prototypes, but retain a softer more irregular style. Green and brown glazes were used often with simple incised or stamped decoration.

The tea ceremony and its influence

The tea ceremony was introduced to Japan by Zen Buddhist monks in about the 12th century. Powdered green tea, drunk as a sort of broth, had been used as an aid to meditation by monks in China during the 11th century, and this way of drinking tea became widespread in Japan. By the 15th century it had become a secular ceremony, patronized by the military aristocracy of Kyoto; it was often combined with lavish banquets, or was sometimes conducted in a more literary and artistic atmosphere. Scholarly Zen monks advised their patrons on the aesthetics of the ceremony and the teawares themselves; gradually the rules became formal and codified. Initially fine Chinese ceramics such as celadons and the brown-black tenmoku teabowls of the Song dynasty were used.

An Arita blue and white ewer with Dutch silver mounts, c. 1660–1680. The production of blue and white porcelain in China was severely disrupted by the wars at the fall of the Ming Dynasty and the Dutch traders turned to Japan for porcelains which copied the styles of the Chinese "transitional" wares of the 1640s and 1650s.

A large Imari charger, late 17th century. Imari wares, characterized by the use of underglaze blue, iron red and gold, were exported through the port of Imari to the West where they achieved a popularity second only to that of Kakiemon.

With the rise of the merchant class to a position of considerable wealth and power, more people were drawn to the tea ceremony and it began to evolve into its present form. The tea-room was set apart for the ceremony itself and there was a developing connoisseurship of the utensils. There was a shortage of the traditional Chinese teawares, so locally produced ceramics were introduced to the ceremony. Great tea masters such as Sen no Rikyu (1521–1591) were the arbiters of taste; a spirit of simplicity and frugality predominated. Amongst the first Japanese ceramics to be accepted for the tea ceremony were the coarse-grained wares of Shigaraki and Bizen and the irregularly formed Iga wares. The peasant rice bowls of Korea were also admired for their primitive dignity; their

An Arita blue and white dish of Kakiemon type, early 18th century. The finest Arita blue and white porcelains used a similar body to that of the polychrome Kakiemon wares.

imperfections were seen as virtues.

Tea masters ordered wares for the tea ceremony from kilns in Seto and Mino provinces – such as the delicate yellow Seto wares, and the Shino wares with their thick bubbly glazes. Also notable are the Oribe wares, named after the tea master Futura Oribe, which combine rich green glazes with bold and striking geometric designs in asymmetric compositions. Under the direction of tea master Rikyu, the potter Chojiro Raku developed the ware that bears his family name. The manufacture of raku teabowls required little equipment; the ware was fired to a fairly low temperature before being plunged into warm water, which gave it a crackled glaze. The coarse body was moulded by hand and the simple bubbly glaze gave it a pleasing irregularity. The softness of the bowl meant that the surface would slowly alter through prolonged use, adding to its interest. At Kyoto a quite different style of tea bowl was developed by the artist-potter Nonomura Ninsei in the second half of the 17th century. These highly ornate wares used a range of brilliant colours and much gilding, without losing a certain rusticity in their potting. Kyoto continued to be a major producer of teawares. A tradition of artist potters flourished and continues to this day, often passing from father to son through many generations.

Porcelain production

After the invasion of Korea by Toyotomi Hideyoshi at the end of the 16th century, the returning armies brought back many Korean potters, who had an important effect on Japanese ceramics. Initially the Korean potters produced the rough Karatsu wares, much admired by the tea masters, but early in the 17th century they made the first porcelains in Japan. These porcelains were made from fairly poor clays and were simply decorated in a greyish cobalt blue. By the middle of the 17th century the centre of Arita produced fine blue and white wares, showing Chinese influence. The use of enamels was also introduced at this time.

In 1641 the Dutch East India Company set up a trading post on Deshima Island, near to the Arita kilns. The Dutch had been trading in Chinese porcelain, but with the

A Kakiemon figure of a young man, c. 1660–70. This figure came from Burghley House, Lincolnshire, where in 1688 the 5th Earl of Exeter had an inventory made of the contents of the house. This figure was described as an "Indian Queen" as people in England had little idea of the countries of the East. It is one of the earliest Kakiemon pieces recorded in England.

collapse of the Ming dynasty in 1644, porcelain production was disrupted. An alternative source of porcelain could now be found in Japan, and the Dutch exported vast quantities to Europe during the 17th and early 18th centuries. Because of the extent of this trade, in the West we are more familiar with Japanese porcelain than with the earlier teaware tradition.

There is still much uncertainty about the correct attribution of Japanese porcelains to particular groups and kilns. Certain decorative groups of the 17th and 18th centuries have been given names that may, strictly speaking, be incorrect, but they have become too useful and widespread to be rejected entirely. The blue and white wares are known as Arita wares, the pieces decorated in underglaze blue, iron-red and gold are called Imari and the distinctive group enamelled in green, blue, red and yellow are termed Kakiemon wares. There are many subgroups, however, and excavations have shown that different types of ware were often made at the same kilns.

The Arita blue and white wares are frequently close copies of Chinese late Ming dynasty or Transitional prototypes (see page 52), but in a softer, slightly greyish blue. Many shapes were also derived from European metalware and pottery. Large chargers and massive garnitures of vases were made in both blue and white and in the Imari palette, intended for the great European houses of the period where they can frequently still be seen. The most sought after wares, both today and in the 18th century, are the Kakiemon wares. Tradition says that Sakaida Kakiemon learnt the skills of enamelling from a Chinese potter in Nagasaki in the early 17th century but the style as we know it seems to have begun to develop in the 1660s.

What we know as Kakiemon wares were not made exclusively at the Kakiemon kiln. Several Arita kilns made the fine milky-white body known as "negoshide" which characterizes the finest Kakiemon wares; it is known that the Kakiemon kiln also produced blue and white wares. Vases, teawares and a wide range of human and animal figures are found in Kakiemon porcelain; the sheer brilliance of the palette ensured its popularity, and spawned many imitators in Japan and Europe.

Arita, Imari and Kakiemon wares were all made for export. Porcelain was also made for the domestic market. The rare Kutani dishes were much prized in Japan – they are densely decorated in thick green, yellow, aubergine and blue enamels on a rather rough body. Especially fine are the Nabeshima wares made just north of Arita for the Nabeshima *daimyo* (lord) of Hizen; they were made with extraordinary precision, the technique being based on that of the contemporary Chinese *doucai* (clashing colours) wares. The brilliant enamels are used in designs of remarkable inventiveness.

A Ko-Kutani dish, late 17th century. This rare group of porcelains, made for the domestic market, is notable for its richly inventive designs and bold use of colour.

The later 18th century was a period of decline in Japanese ceramics, as an inward-looking regime forbade contact with the outside world. The Hirado kilns started in the 18th century and by the early 19th century were producing both white and blue and white porcelains of a very high degree of technical excellence. The Meiji restoration of 1868 vigorously reopened trade with the west, and Hirado ware was then exported to new markets. The later 19th century saw a great expansion of the industry; Satsuma and other wares flooded into Western markets. At their best these were of very good quality, but mostly they were crudely executed and vulgar in design. Certain potters, such as Miyagawa Kozan, were active in adopting new techniques and created a wide variety of very original works of high quality.

The ceramic tradition continues to thrive in Japan, where the great artist potters are held in the highest regard – some are even classified as "living national treasures" – and command truly stupendous prices.

INFLUENCES

1

*A print by Toshikata (**1**), c. 1900, of ladies preparing for the tea ceremony. The evolving aesthetic of the tea ceremony exercised an overwhelming influence on Japanese ceramics.*

*An Ido type Korean teabowl (**2**), 16th century. The simple peasant pottery of Korea was highly revered for its noble austerity; this example is believed to have been used by the teamaster Oda Uraku (1547–1621).*

*The great ceramic tradition of China inevitably had enormous influence in Japan, particularly in the late Ming period of the 16th and early 17th centuries. The Swatow dish (**3**), c. 1600, in* kinrande *style and the enamelled dish (**5**), c. 1620–1640, would have been made especially for export to Japan. Although the blue and white dish (**4**) of the Wanli period (1573–1619) would have been made for the Chinese domestic market, it would also have found favour in Japan.*

2

3

4

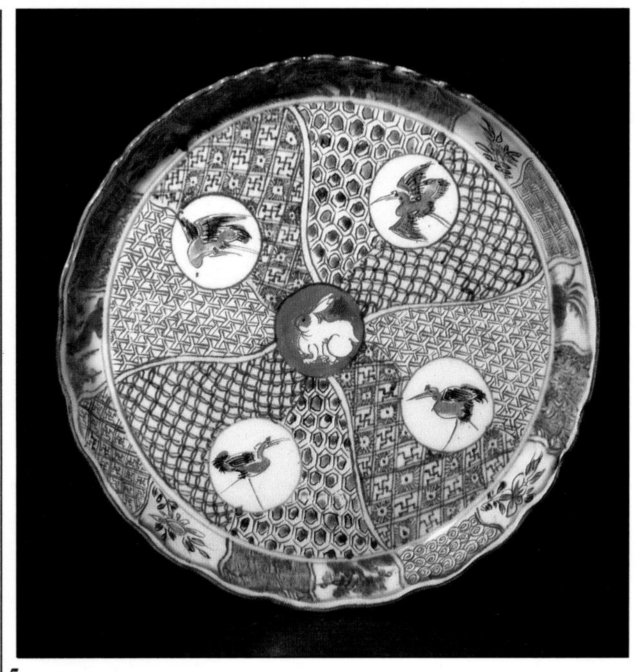

5

*Irises (**6**), by Ogata Korin, watercolour on a screen, c. 1700. Korin was one of Japan's greatest decorative artists, whose influence can be seen on the teawares of the Kyoto school.*

6

EARLY WARES

1

*Jomon pottery jar, probably for storing food (**1**), middle Jomon period, c. 3000–2000 BC. The Jomon period extended over thousands of years; the middle period was the high point for pottery. These wares are characterized by the often bizarre and convoluted decoration formed in high relief.*

*A Haniwa figure of a warrior (**2**) in Keiko armour, 6th century AD. The curiously arresting Haniwa figures of warriors and horses come from the funerary tumuli of the period. Comparison with contemporary iron armour shows that it is faithfully reproduced in this figure.*

2

*A Sue ware jar (**3**), 6th century AD. The techniques for producing these high-fired stonewares were introduced from Korea. The irregular areas of glaze result from chance falls of wood ash in the reducing atmosphere of the kilns.*

*A Seto ware jar (**4**), 14th century AD. Influenced by the Chinese ceramics of the Song Dynasty, these early stonewares were much admired by later Japanese potters.*

3

4

TEA CERAMICS

1

2

A Shigaraki tsubo *or jar* (**1**), *16th–17th century. The rough qualities of Shigaraki ware with its irregular glaze splashes were admired even before they were accepted as being suitable for the tea ceremony. Tsubo were used for the storage of leaf tea.*

The base of an Oribe chawan *(teabowl)* (**2**), *17th century. The Japanese teamasters appreciated the rugged austerity of these wares; the studied lack of refinement was considered conducive to the dignity of the tea ceremony.*

3

An Oribe stoneware squared bowl (**3**), late 16th century. The startling boldness of the design is typical of Oribe stoneware.

4

5

A black-glazed raku earthenware chawan *or teabowl (**4**), 17th century. These low-fired wares were simply moulded by hand and much of their appeal lies in the thick bubbly glaze that makes them pleasing to hold.*

*A Shino ware bowl (**5**), late 16th/early 17th century, with typical indistinct but expressive painted decoration.*

BLUE AND WHITE

1

2

*The large Arita charger, (**2**) c. 1670 bears the VOC monogram for the "Vereenigde Oostindische Compagnie", the Dutch East India Company, and was probably intended for use in one of their overseas offices. When supplies of porcelain became hard to obtain from China the Dutch traders encouraged the Arita potters to produce large quantities of blue and white porcelain. The beaker vase (**4**) of the late 17th century would have formed part of a garniture of vases in one of the great European houses or palaces of the period.*

4

3

*This dish (**1**) is an example of the finest Arita blue and white from the Kakiemon kiln in the late 17th century.*

*This rare Arita teapot (**3**) of the early 18th century has a design derived from a European print, with a very European interpretation of an Oriental scene.*

KAKIEMON

*This rare late 17th century model of a tiger (**1**) makes little attempt to be naturalistic but the naïvety of the modelling is complemented by the vitality and humour of the painting.*

2

1

THE KAKIEMON PALETTE

The brilliant Kakiemon palette was at its best when combined with the fine milky-white porcelain known as negoshide, developed in the late 17th century. Kakiemon porcelains were made primarily for export to the West where they were prized above all others and are still to be found in the great houses of the period. The finest collection was that at Dresden, assembled by the Elector Frederick Augustus I of Saxony who founded the Meissen factory.

Many of the early European factories copied Kakiemon wares; the designs left much of the surface in white which allowed the factories to display the quality of their porcelain to the fullest advantage. After the 1750s the fashion for these wares declined and the European copies which were then made, notably in the English factories such as Worcester, ceased to be the almost exact replicas of the earlier period. There was an increasing tendency to overformalize and complicate the designs whilst retaining elements of the Kakiemon palette.

*The design of this octagonal dish (**2**) c. 1680, became known in England as "Hob in the well" and was copied at Meissen and Chelsea. It depicts an old Chinese story in which the young Sze-ma Kwang saves his friend from drowning by smashing the water jar into which he had fallen.*

3

4

*This popular shape (**3**), c. 1680, is sometimes known as a "Hampton Court" vase. Queen Mary II brought the fashion for Kakiemon ware from Holland and amassed a large collection at Hampton Court where some of it can still be seen today. Vases of this type were never really intended to perform any practical function but were used solely for display.*

*This design (**4**) c. 1680, of pheasants (sometimes known as hobo birds) admirably displays the characteristic asymmetry of the Kakiemon designs which must have been so startling to the Europeans of the period. This design became known as the "Joshua Reynolds" pattern on Chelsea and Worcester porcelain.*

IMARI

*A pair of triple gourd vases (**1**), late 17th century. These vases, made in Arita province and exported through the port of Imari, were destined for the great houses of Europe. The Dutch East India Company dominated the trade which declined in the early 18th century.*

1

2

*This large and striking charger (**2**) of the late 17th century is painted with eagles, a popular subject on this class of porcelain. The characteristic Imari palette of underglaze blue, iron-red and gilt has been sparingly accentuated with green enamel.*

*The charger (**3**), c. 1700, displays a palette which, although not typical of the style, may still loosely be called Imari.*

3

4

*A pair of Imari cockerels (**4**), c. 1700. Whilst figures such as these are comparatively rare, the use of smaller versions to decorate the finials of vase covers was widespread.*

ENAMELLED WARES

1

*Kutani dish (**1**), late 17th century. These rare wares were made from a rather coarse porcelain exclusively for the domestic market; the rich palette and matt glaze is characteristic of the group. No two examples of Kutani ware are the same and this striking design typifies the inventiveness of the painters.*

2

*This arresting figure of an eagle (**2**), late 17th century, is almost life-size. It inspired the modeller Johann Kirchner at Meissen to produce very close copies in the 1730s.*

3

*A Nabeshima dish (**3**), early 18th century. The "official" kiln of the Nabeshima* daimyo, *or lord, of Hizen produced some of the finest porcelains, painstakingly enamelled in a technique based on the* doucai *(clashing colours) porcelains of China. The standard of potting and painting was always extremely high.*

*The Arita jar (**5**), c. 1680, is of a type made for the domestic market: the boldness of the decoration is typical of the period.*

*The Arita puppy (**4**), c. 1670–80 would have been one of a pair, made for export to the West. In spite of their lack of sophistication these models exhibit a great deal of character and liveliness.*

4

5

LATER WARES

1

2

*Blue and white vase (**1**), by Makuzu Kozan, Meiji period. The technical mastery of the Japanese potters at the end of the 19th and beginning of the 20th century ensured a considerable demand for their work in the West.*

*An oviform vase (**2**) in the Satsuma style, Yabu Meizan workshop, Meiji period. Much Satsuma pottery of poor quality was made and exported; the finer works, of which this is an example, were very accomplished.*

*Two serving dishes (**3**) by Kitaoji Rosanjin (1883–1959). Rosanjin was one of the most highly regarded modern potters, working predominantly in the traditional styles such as Shino and Oribe.*

3

4

*A vase painted with a tethered hawk on a perch (**6**), Makuzu Kozan, Meiji period. Based on a famous painting, this vase exhibits the astonishing virtuosity of Kozan's work.*

5

*Porcelain bowl (**4**) by Shinobu Kawase (born 1950). The third generation of a distinguished line of potters, Shinobu has been greatly influenced by the Chinese Song dynasty tradition of celadons.*

*A Satsuma jar (**5**), Meiji period. Many workshops worked in this style, using enamels on a cream ground.*

6

A slip-painted bowl, c. AD 1000, Nishapur, east Persia.

CHAPTER · FOUR

THE ISLAMIC WORLD

"I watch'd the Potter thumping his wet clay:
And with its all obliterated tongue
It murmer'd – 'Gently, Brother, gently, pray!"

EDWARD FITZGERALD –
THE RUBÁIYAT OF OMAR KHAYYÁM

THE ISLAMIC WORLD

Ceramics, in the area covered by the Islamic empire, had not historically played a very important part in the decorative arts but the basic technology of alkaline and lead glazes and fritware (see Introduction) did exist. There were various factors that encouraged a sophisticated industry to develop – the wealth of the Abbassid capital of Baghdad, which by the 9th century was probably the largest city in the world, and the Islamic prohibition of gold and silver vessels. Perhaps the greatest spur to the production and refinement of Islamic pottery was the import of ceramics from Tang China. Nothing outside China could rival the well-potted porcellaneous white stonewares and the coloured and splashed earthenwares. These were imported along the silk route and by sea, and have been excavated at sites in the Middle East. Chinese porcelain continued to inspire Islamic potters in later centuries and was often a decisive influence on form and decoration. The very high regard in which Chinese porcelains were held is testified to by the great collections that still exist in the Topkapi Serail at Istanbul and the Ardebil Shrine in Iran, presented by Shah Abbas in 1611. They both contain large quantities of blue and white porcelain of the finest Imperial quality, as well as celadons and other lesser wares.

In the *Hadith*, a work recording the sayings and practices of the Prophet Mohammed, the depiction of human and animal forms was forbidden in all the decorative arts. Although this prohibition was widely disregarded, it encouraged other forms of decoration, such as the wide repertory of geometric designs and calligraphy that played an important part in Islamic pottery.

Early Islamic wares

The Mesopotamian wares of the 9th and 10th centuries are the first important Islamic pottery. Two major innovations occurred which influenced enormously the ceramic traditions of both the Islamic empire and Europe. The first was the reinvention of the tin glaze, which gives a good white surface to paint on. This may well have been in response to the white Chinese Tang wares. A limited early palette of blue and green was used with simple calligraphic designs, palmettes or the occasional animal. At this stage potters found it difficult to control the flow of colours in the glaze, which gives the wares a characteristic "blotting paper" effect. The tin glaze technique was later introduced to Europe where it became the mainstay of the pottery tradition. The second important Mesopotamian innovation was the use of metallic lustres on pottery, probably developed from the techniques of Egyptian glass. Lustres were used extensively throughout the Islamic world; they were introduced to Spain by the Moors, leading to the Hispano-Moresque ceramic tradition, which in turn directly influenced the lustred wares of Renaissance Italy.

A dish with moulded decoration under a lead glaze with areas stained green and partially lustred; Mesopotamia, 9th century. One of the earliest Islamic fine wares, this dish was formed by pressing the clay over a mould which left the design in relief. The lustre, which is somewhat degraded, was probably intended to imitate metal vessels.

Contemporary with the early Mesopotamian wares are the slip decorated wares of Persia and Afghanistan, associated with Nishapur and Samarkand. Although these cities were on the trade routes to the East, little Chinese influence is discernible on the wares except for the splashed and mottled-glazed bowls which recall the Tang dynasty three-colour or *sancai* wares. The finely grained buff or red earthenware bodies of the Persian wares were covered with a white clay slip and then with a clear colourless glaze. Some of the most striking pieces, often linked with Nishapur, have bold calligraphic borders in the Kufic script. An important discovery was that colouring pigments when mixed with fine white clay did not run under the glaze, which allowed the potters a more defined style of decoration.

By the beginning of the 10th century the Abbassid caliphs had lost control of their vast empire, and local potentates and invaders founded their own dynasties. Islam was never again to become a completely unified political force – yet in spite of this fragmentation the Islamic arts continued to flourish.

Artists from Mesopotamia had settled in Fustat, Old Cairo, in the 9th century and under the Fatimid Dynasty, founded in AD 969, Egypt became the cultural centre of

An Iznik dish, c. 1540–1550. This dish of the group usually known as "Damascus ware", already displays the stylized floral designs that distinguish much later Iznik pottery.

Islam. The most notable pottery produced under the Fatimids were the lustre wares; these were of a coarser clay body than their Mesopotamian predecessors, but during the 11th and early 12th centuries the painting attained a high level of artistry. The range of subjects was wide and included vivid depictions of humans, animals and birds – sometimes Christian influence is evident – a Coptic priest is shown on one particular pot held in the Victoria and Albert Museum. After the fall of the Fatimids in 1171, pottery declined and some potters moved east to Mesopotamia and Persia and west to southern Spain, taking their skills with them and invigorating the ceramic traditions of these areas.

Seljuq wares

The Seljuq Turks from Central Asia conquered Persia, Iraq, Syria and Asia Minor in the 11th century and were converted to Islam in the process. Their arrival inaugurated what is considered to be the great classical age of Islamic art. Pottery, particularly at Rayy and Kashan in Persia, rose to new heights. Inspiration was again derived from China; in an attempt to emulate the ivory-white Ding wares and Yingqing porcelain a vitrified fritware was developed that had a fine white body and was slightly translucent when thinly potted. These relatively rare Persian white wares of the 12th century, with their finely carved decoration, are amongst the most beautiful of all Persian pottery. The alkaline glazes of the frit-wares produced very striking monochrome glaze colours notably the brilliant turquoise derived from copper and a rich cobalt blue.

The inventive Seljuq potters used a range of different techniques when decorating their pottery. The *laqabi* technique solved the problem of coloured glazes running together: areas of different colour were separated by raised or incised lines, rather in the manner of cloisonné enamel. The technique lent itself to the production of dishes, which were often boldly decorated with birds and plants. The Seljuqs also used the old technique of carving a design through a dark slip before glazing, a group of ware known as "silhouette-painted". A method of underglaze painting in black under a clear or turquoise glaze produced some striking results. A very successful innovation was the development of overglaze decoration known as *minai* ware; here the decoration was painted onto the glazed body and refired at a lower temperature. The decoration closely resembles the miniature painting of illuminated manuscripts and often depicts delightful court and hunting scenes and stories from Persian legends. The good range of colours and the addition of gilding make these amongst the most luxurious of the 12th and 13th century wares. A related technique to *minai* was that of *lajvardina*, in which a more limited range of colours was applied to a dark cobalt blue background; the decoration is generally confined to geometric designs.

Lustre wares

In the 12th and 13th centuries at least five important pottery centres were producing lustre wares in Persia and Mesopotamia – pre-eminent amongst these were Kashan and Rayy. A wide range of vessels and figures were made, and tiles for the decoration of mosques formed an important part of the production. The quality of the painting was often excellent; amongst the most successful styles of decoration are the courtly ladies reminiscent of the "moon-faced beauties" of early Persian poetry. The Mongol invasions of the 13th century brought pottery production to an end in many centres, but at Kashan the industry continued to thrive under the new rulers well into the 14th century. Some new developments occurred there, such as copies of the large Chinese celadon dishes that were becoming popular at the time.

With the exception of fine architectural tiles for mosques and mausolea, pottery in Persia declined in the 15th and the greater part of the 16th century. The production of lustre wares ceased almost completely except for some rather inferior tombstones, which show that the techniques were not entirely lost. The few wares that can be dated to the 15th and 16th centuries, like the Kubachi dishes, tend to show a strong Chinese influence.

A vase painted in lustre on an opaque white glaze, Persian, late 12th century. The "moon-faced beauties" are typical of Persian art and were often found on wares in the Rayy "monumental style".

An Iznik dish, c. 1520–1530. The early blue and white Iznik wares were strongly influenced by the Chinese porcelain of the 15th century.

Safavid pottery

The Safavid dynasty (1499–1736) was a period of political stability and economic growth and a considerable revival of pottery occurred throughout this time. Towards the end of the 16th century an independent Safavid style developed, the wares often being decorated with birds and foliage. The Chinese influence continued, however, particularly in the blue and white wares. The blue and white wares associated with Meshed are of particularly good quality, with the decoration outlined in black. They are usually, but not always, in the Chinese late Ming style and often bear pseudo-Chinese marks. The Safavid lustre wares are more often decorated in the Persian style; attractive ruby and brown lustres were used in conjunction with blue and yellow to create some fine pottery. Amongst the finest Safavid wares are a wide range of monochromes. Of these, the white Gombroon wares are extremely beautiful – finely potted with incised decoration and very translucent they are reminiscent of the earlier white wares developed by the Seljuqs. Other fine quality monochrome glazes, yellow, brown, red, blue and celadon were used sometimes with relief decoration. After 1700 there was a sharp decline in quality as cheaper imports from China undercut the local producers and little further pottery of any note was made in Persia.

Ottoman wares

Of all Islamic pottery, perhaps the best loved by Westerners are the Iznik wares made under the Ottomans in Turkey. The brilliant colours and designs influenced the maiolica potters of Italy and were one of the major sources of inspiration for the Arts and Crafts potters at the end of the 19th century.

The first wares made under the Ottomans are the blue and white wares known as the "Abraham of Kutahya" group, which were produced under courtly patronage from about 1490 for a period of around thirty years. They display a high level of technical proficiency, and are painted with intricate arabesques and scrolls showing some Chinese influence. Also in blue and white are the "Golden Horn" type wares – these have a very distinctive style of decoration, formed as tight spiral floral sprays. These spirals are also found on the "Tugra", the calligraphic imperial monograms of Sultan Suleiman the Magnificent (1520–66) suggesting that the court exerted a strong influence on pottery production.

The Iznik wares which were most influential in the West were first made about 1520; by the middle of the century a highly developed palette had been acquired. These mid 15th century pots are known as the "Damascus" type wares and their designs were probably by court artists, as similar designs can be found on courtly carpets and textiles. The slightly later wares have in addition a fine red colour, and are painted with brilliant designs of carnations, tulips and roses. The tradition continued throughout the 17th century, but the decline in quality was swift and the later wares are a poor shadow of those from the 16th century.

Armenian Christian potters were responsible for a further group of wares in Turkey, which reached their finest production in the 18th century. Whilst they never achieved the grandeur of the courtly Iznik wares, a fine white body was used to good effect and at its best the painting is most attractive. Smaller pieces such as tea and coffee services predominate and the shapes often display European influence.

Islamic potters never made true porcelain, but their rich legacy of designs and techniques, absorbing many influences whilst creating a uniquely Islamic flavour, can well hold its own amongst the world's ceramic traditions.

INFLUENCES

1

2

*A 14th century inlaid brass ewer (**2**) from Egypt or Syria, and a 14th century Syrian lustre ewer (**1**) of similar form. Ceramic and metal wares often fulfil the same functions and so it is not surprising to find a cross fertilization of ideas in both form and decoration that frequently disregard the methods of construction and natural properties of the different materials.*

3

*A Chinese blue and white dish (**3**) of the Yuan dynasty, mid 14th century. Chinese ceramics were imported in large quantities to the Islamic world and were held in the highest regard. From the 7th century Chinese wares were a constant stimulus to the Islamic potters.*

*A Syrian or Egyptian candlestick of beaten brass inlaid with silver (**4**) dated 681 (AD 1282). The metallic lustre used by potters throughout the Middle East was used to simulate the inlaid decoration on metal wares.*

4

5

6

7

Zafarnāma (**5**) (AD 1548), from the Life of Timur. The figurative decoration of Iranian ceramics recalls the works of the miniaturists.

*A Chinese Tang dynasty white porcelain bowl (**6**), 9th century. Examples of these, the earliest porcelains, have been excavated at Samarra where they clearly influenced the potters of Mesopotamia.*

*A late 14th century mosque lamp (**7**). Potters learnt many of the techniques of enamelling and lustre from the glassmakers.*

A Chinese sancai *or three-coloured ware bottle (**8**), Tang dynasty, 8th century. The carefully controlled and splashed lead glazes became an important part of the repertory of decorative techniques.*

8

ABBASSID

*A large jar (**1**) with a turquoise blue glaze, Mesopotamia, 8th–9th century. Storage vessels of this type were made in the pre-Islamic period and similar jars are still made today in remote parts of Turkey.*

*A tin-glazed bowl (**2**), painted in green and cobalt blue with an inscription in the centre, Mesopotamia, 9th–10th century. The addition of tin oxide to the glaze during this period gave a white surface suitable for painted decoration.*

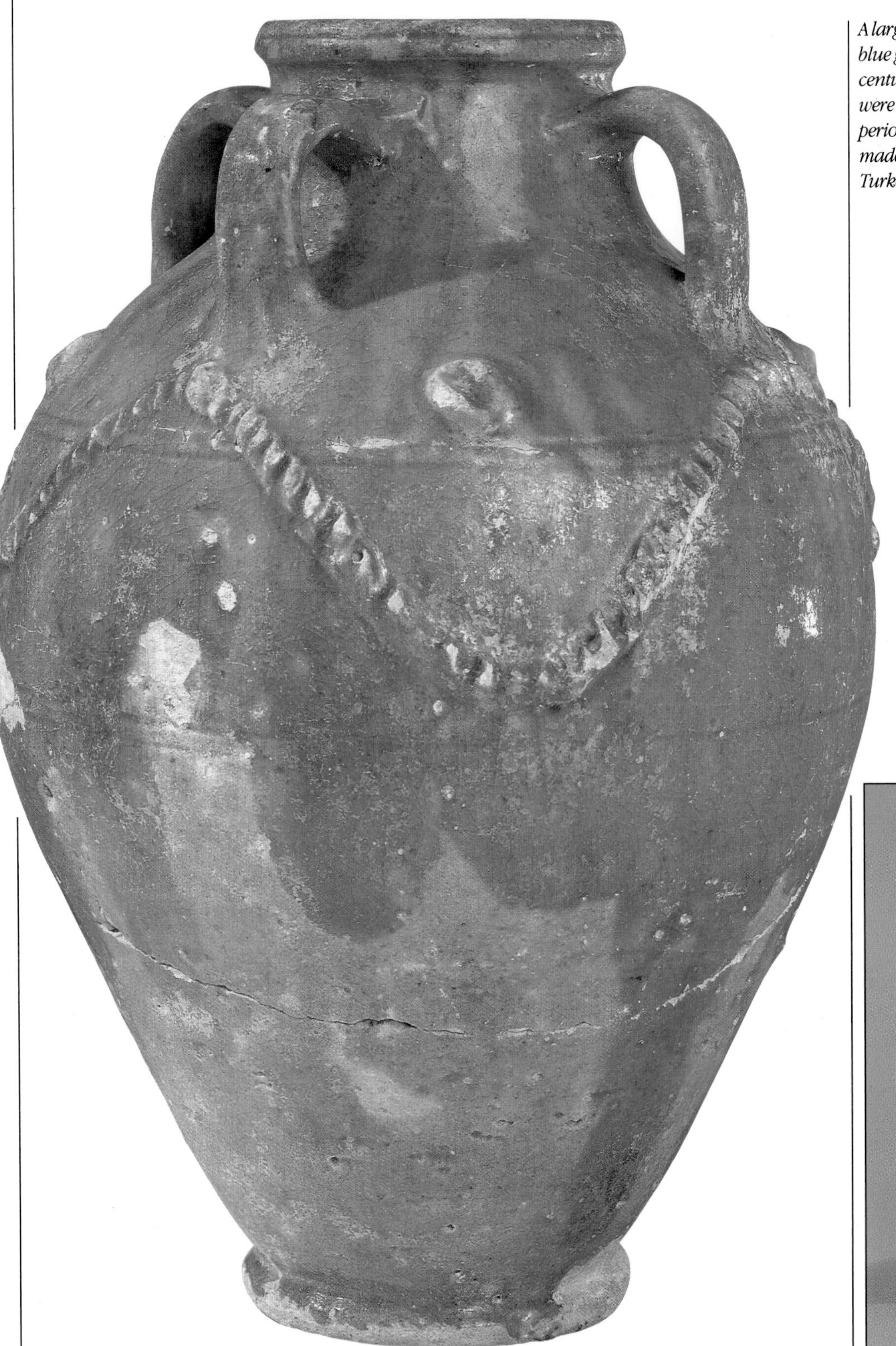

1

2

3

*A bowl painted in lustre on a tin glaze (**3**), Mesopotamia, 10th century. The lustre technique evolved from the skills of Egyptian glass decoration. The seated figure shows the influence of the art of central Asia.*

4

*A bowl decorated in two shades of lustre (**4**), Mesopotamia, 9th century. The early lustre wares employed a range of different coloured lustres, whereas the slightly later wares tended to use only one.*

5

*A dish painted in cobalt blue on a tin glaze (**5**), Mesopotamia, 9th century. The fish design is unusual and ambitious for this type of ware.*

EARLY IRAN

1

*A Nishapur bowl (**1**) from east Iran, 10th century, painted with a ewer and palmettes. The striking designs of these early wares were painted in coloured slips under a clear glaze. In this way the design would not run into the glaze and could be controlled.*

3

*A champlevé ware bowl (**3**), north-west Iran, 11th–13th century. The decoration on champlevé ware is carved through a thick white slip to reveal the body that was often darkened by painting with manganese brown. The ware was covered with a transparent lead glaze.*

2

*A Nishapur bowl (**2**), 10th century, painted with a band of stylized Kufic script. Amongst the most refined of all Persian pottery, the "epigraphic" ware relies solely on the beauty of the calligraphy for its effect.*

*A dish, Amul ware (**4**), north-west Iran, 11th–13th century. The design was scratched through the slip to reveal the body and further painted in a green that has run slightly into the glaze.*

4

5

*A mug (**5**) painted with a Kufic inscription; "silhouette" ware, Iran, 12th century.*

*A bowl (**6**) painted in slip under a colourless glaze; Samarkand, 9th–10th century. Similar designs were used at a later date on the tiles and textiles of Ottoman Turkey.*

6

SELJUQ IRAN

1

*A Minai ware bowl (**1**), Iran, late 12th century, tin glaze painted blue, turquoise, red and black enamel. This type of decoration, showing complex arabesques, is associated with a group of Minai wares which were made by the same potters who made lustre wares in the Rayy style. The well potted Minai wares are amongst the most luxurious of the period.*

*An albarello or drug jar (**2**), Syria, 14th century, painted in lustre on a clear blue glaze. Drug jars were used by apothecaries for the storage of spices and drugs; the shape became firmly established in Renaissance Europe.*

*A cat (**3**), painted in lustre on a white glaze; Iran, 12th–13th century.*

2

3

*A 13th century Syrian bowl (**4**), incised and stained with glazes. The sgraffito technique was used with particular verve and dynamism by the Syrian potters.*

*A Lajvardina ware bowl (**5**), Iran, first half 14th century. Lajvardina ware superseded the Minai wares after the Mongol invasion of Iran. The technique was similar to Minai but the decoration was on a dark blue or turquoise ground and gilding was introduced.*

4

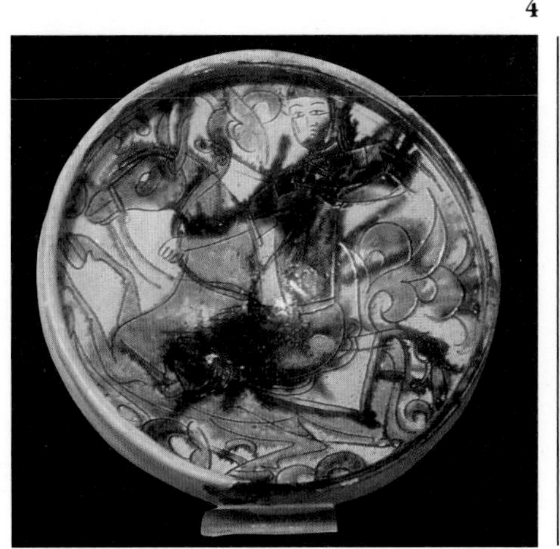

5

OTTOMAN

1

2

*An Iznik bottle vase (**2**) and a particularly fine jug (**4**), both 1560–1580. The rich palette and beautiful designs of Iznik pottery excited both admiration and imitation in Europe and were later to influence the Art potters of the late 19th century.*

3

*An Iznik tile panel (**1**) c. 1580. The Ottoman Turks continued the tradition of their Seljuq predecessors by decorating the inner walls of buildings with magnificent coloured tiles.*

*An Iznik blue and white dish (**3**), mid 16th century. Made in imitation of Ming porcelain, it still retains a distinctly Ottoman flavour.*

4

1

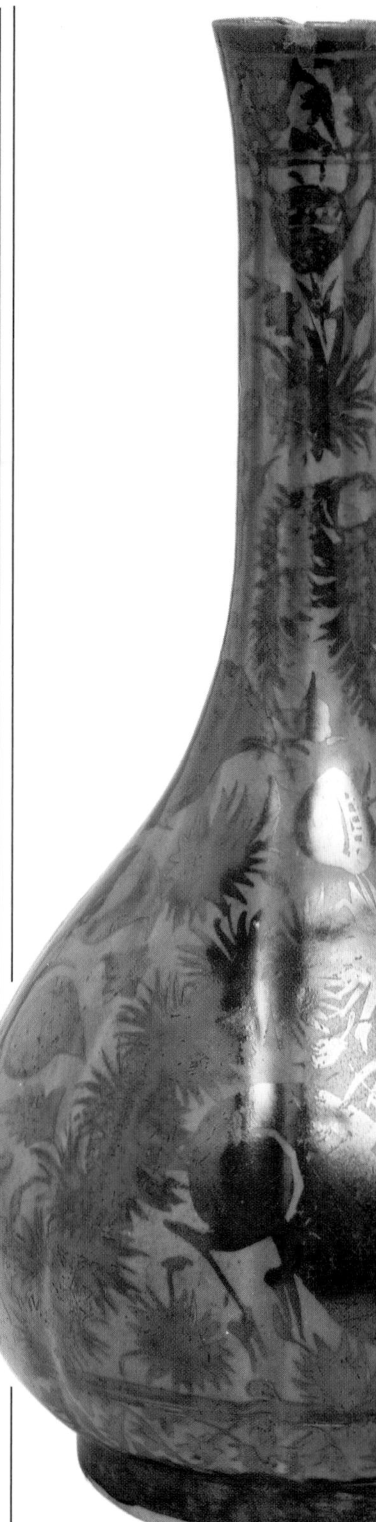

2

A blue and black painted elephant Kendi *or ewer (**1**), first half of the 17th century. This is a very close copy of a Chinese late Ming original of a type that was widely exported throughout South East Asia and the Middle East.*

*A bottle painted in lustre on a blue ground (**2**), second half of the 17th century. The 17th century saw a revival of the lustre technique that had been largely neglected since its heyday in the 12th and 13th centuries.*

3

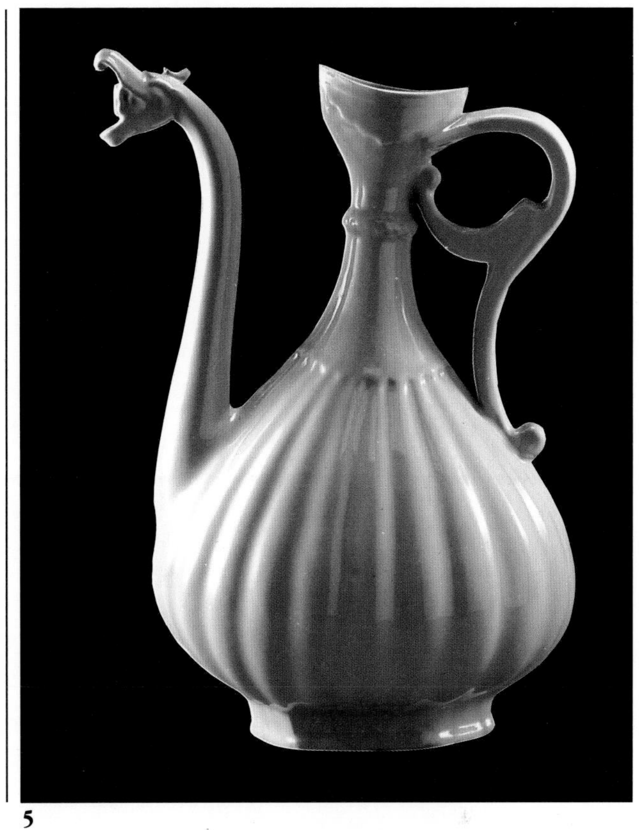

5

4

*Although the Chinese influence is rarely completely absent in the wares of Safavid Iran, the strong indigenous styles of decoration and shapes impart an unmistakably Iranian flavour. The blue and white dish (**3**) of the late 17th or early 18th century, is painted with the birds and flowers that are a particular feature of Safavid art.*

*A "Gombroon" ware bowl painted in blue and black (**4**), early 18th century. This type is a revival of the 12th century white wares that were themselves inspired by the Chinese Song dynasty monochromes.*

*A pale blue-glazed ewer (**5**), 17th century, the form is a pastiche of a contemporary brass ewer possibly intended for the Mughal market.*

A maiolica dish, perhaps from Faenza, c. 1510–20. The painting is of the death of the Virgin, taken from the engraving by Israel Van Meckenem after Martin Schongauer.

CHAPTER · FIVE

EUROPEAN POTTERY

"John Udy of Luscillion
his tin was so fine
it gliderd this punch bowl
and made it to shine
pray fill it with punch
lett the tinners sit round
they never will budge
till the bottom they sound."

FROM AN ENGLISH
DELFTWARE PUNCHBOWL, 1731

EUROPEAN POTTERY

Each European country has its own tradition of pottery, and these are all closely interrelated. Potters could and did travel freely, taking the secrets of their craft to new countries where patronage or markets could afford them a living. The production of earthenware requires only a low level of technology and small potteries could be set up wherever suitable clay was to be found. Small potteries supplying local needs are still found throughout Europe and they have in the past produced rustic masterpieces (for example the Staffordshire slipwares and mid-European peasant pottery) as well as a wide variety of domestic wares, bricks, tiles and flowerpots. In contrast, there were also great workshops producing commissions for princely patrons, supreme examples of these being the maiolica workshops of the Italian Renaissance. The Italian maiolica of the first half of the 16th century, the greatest of all European pottery, has long been sought after and collected by connoisseurs.

For much of its history, pottery was considered inferior to work done in silver and gold; having no intrinsic value, it could compete purely in terms of its forms and decoration. The long, arduous journey along the silk routes brought a few rare and expensive porcelains from China to Europe through the great trading city of Venice; these treasured wares inspired a greater refinement of pottery but it was many years before porcelain could be made in Europe.

Large dish, Fontana workshop Urbino, c. 1561. The middle is painted with the meeting of Abraham and Melchisedec, the reverse with the arms of Cardinal Inigo d'Avalos e Aquino. The border pattern of grotesques derives from antique models popularized by Raphael.

Tin-glazed pottery

European pottery was dominated by the tin-glaze tradition. The earthenware clay bodies used in Europe varied in colour from brick-red to off-white and were in themselves an unsatisfactory background for painted decoration. The addition of tin oxide to the glaze produced an opaque white surface which was most suitable for painting. Tin oxide was expensive so often only the front of dishes and the outside of vessels were glazed in this way. The best tin glazes produce an even, brilliant white surface and were usually used on light-coloured bodies. Confusion arises over the different names used to describe tin-glazed wares from different countries. Maiolica, Faience and Delft are all technically the same – earthenware covered with a tin glaze. Maiolica is used to describe the early tin-glazed wares, particularly those of Italy. Faience is used to describe those wares made in France, Germany and Scandinavia. Delft is the name for the wares of Holland and Britain.

Tin-glazed pottery was introduced to Europe by the Moorish invaders of Spain. Initially the wares were imported from North Africa and Mesopotamia, but in the 13th century tin-glazed and lustred pottery was produced in Spain. This was the beginning of the Hispano-Moresque tradition that continued for many years. In the 15th century the Hispano-Moresque wares reached their greatest maturity; typically the pieces were large chargers and drug jars painted with bands of oak leaves, or heraldic shields. The coppery lustre was often augmented with deep cobalt blue decoration. The backs of the large chargers were frequently painted in a powerful and free style with eagles or beasts. These fine wares were exported to England, France and Italy, but the quality declined in the 16th and 17th centuries.

Traders from the Balearic island of Majorca brought the lustred Hispano-Moresque wares to Italy, which is where the name corrupted to Maiolica. The term was later used to describe all Italian tin-glazed wares. Tin-glazed pottery was also introduced to Italy from the Islamic world although the earliest types, associated with the town of Orvieto, tend to be decorated in the Gothic style of ornamentation. These 13th and 14th century wares have mostly been excavated as they have rarely survived above

A Niderviller faience teapot, c. 1770 in "décor bois". The trompe l'oeil effect exemplifies the virtuosity displayed by the tin-glaze potters as they struggled to compete with porcelain.

ground. Typically they are decorated in green, manganese purple and black with figures, coats-of-arms or trailing vines. The alborello or drug jar was a form frequently made since monastic pharmacies were important patrons of potters.

Throughout the 15th century the Gothic style was pre-eminent. A rich range of colours was developed. It is worth noting that the great majority of pieces which have survived from the 15th and 16th centuries are the fine courtly wares – these were frequently never intended for use but only for display. The more humble domestic wares were used until they broke. Until about 1500 the production of fine maiolica was dominated by Florence and Faenza. Florence is famous for its two-handled drug jars, often painted in dark blue and showing heraldic beasts on a ground of oakleaves. These powerful pots are entirely Italian in shape but the influence of the Hispano-Moresque wares is still evident in the decoration. It was at Faenza, however, that the greatest developments took place; a very fine and even white glaze was developed which formed a perfect background for both the expanded range of colours and the considerable refinement of the painters' skills.

The early 16th century was the golden age of Italian maiolica. Further centres of production were established in the northern part of Italy: Siena, Casteldurante, Cafaggiolo, Deruta and Venice. Often pots were decorated with copies from the works of artists of the period, such as Mantegna and Raphael. The greatness of maiolica lay essentially in the painted decoration – the forms developed slowly. The painters of the pots began to sign their work, which indicates their increased status. Like other Renaissance artists the maiolica painters considered themselves to be the spiritual heirs of the Greek and Roman civilizations and classical and religious subjects were increasingly used, in what is known as the "Istoriato" style. The great pottery painters of Casteldurante and Urbino such as Nicola Pellipario, Francesco Avelli and Cipriano Piccolpasso are particularly associated with the Istoriato style. By this time, too, there was a finely developed palette of ceramic colours, capable of doing justice to the paintings being copied. The use of lustre (precious metals painted onto the ware) was confined to Gubbio, where ruby-red (derived from copper), silver and gold were used, and at Deruta where fine show dishes were lustred in a coppery gold.

From the middle of the 16th century, the styles of decoration began to change. At Faenza, pots were given white backgrounds with decoration painted only in central medallions. "Grotesques" in the style of Raphael became a popular form of decoration, particularly at Urbino. As the great age of maiolica passed delightful if less aristocratic wares were produced throughout Italy; of particular charm are the dishes of Montelupo, naively

A tin-glaze earthenware jug, Nuremberg, late 17th century. The Nuremberg potters used the "petit feu" or muffle kiln colours to achieve a highly developed palette.

painted with striding soldiers.

Italian maiolica was exported widely throughout Europe, where it was held in high regard. By the early 16th century Italian potters had set up workshops in France and the Netherlands, producing wares in the Italian style, and from Antwerp potters took their craft to England, Germany and Scandinavia. The dissemination of tin-glaze technology laid the foundations for the varied styles that developed throughout Europe in the 17th and 18th centuries. This form of pottery was finally superseded by porcelain and the extensive production of English creamwares.

The emergence of Holland as an economic superpower stimulated a vast pottery industry. In the early 17th century two Portuguese trading vessels, laden with Chinese late Ming blue and white porcelain, were captured by the Dutch and brought to Amsterdam. This Oriental porcelain caused a sensation and ever-increasing quantities were imported by the traders of the Dutch East India Company. These Chinese porcelains were to be the decisive influence on "Delftware". The superb quality of the blue and white Dutch Delftware inspired imitators throughout both Germany and England, where an appealingly idiosyncratic tradition grew up. The expanding middle classes of northern Europe created a large demand for affordable wares and although these rarely matched the splendour of their Italian forebears they can often have a more approachable charm. The English "Delft" potters were rarely patronized by the grandest society but by country squires, yeoman farmers and the growing urban middle classes. The potting was often of a high order and the decoration inventive and varied.

In the 18th century tin-glazed pottery was brought to a very high level of refinement in France. A wide range of low-fired enamels had been developed which were used for a particularly delicate style of decoration. Hundreds of faience manufacturies were set up in France, stimulated by the royal monopoly on the production of porcelain. The quality of painting on the faience of Marseilles and Strasbourg was the equal of all but the finest porcelain decoration.

Lead-glazed pottery

Long before the invention of the tin glaze, lead glazes had been developed in Iran and Iraq and were in widespread use throughout the Roman Empire. The tradition of using these bright clear glazes with their limited palette of green, amber and yellow continued in Byzantium and in Italy during the dark ages. Much of this medieval lead-glazed pottery was for domestic use and consequently little has survived. Handsome, if somewhat primitive, jugs and vessels were made in England and France during the Middle Ages. At their best these pots, often with applied or incised decoration in human or animal forms, are well proportioned and functional and are much admired today.

Lead-glazed pottery was often decorated with the "sgraffito" technique – here the pot is covered with a thin white clay slip which is then incised with a design, revealing the darker clay body underneath. The sgraffito technique was widely used throughout Europe on peasant pottery with delightful results, notably in the German-speaking world and the West Country potteries of England. Later, glazing was brought to a high level of artistry in Italy, and to a lesser extent in France, during the 15th and 16th centuries. The magnificent wares associated with Bologna rival the contemporary maiolica.

Some of the most striking lead-glazed wares are those attributed to Bernard Palissy (c. 1510–1590) and his followers in France. Dishes were often decorated with startlingly realistic lizards, snakes and fish, cast from actual animals. Other wares were finely moulded with a rich range of glazed colours. In Germany, lead-glazed stove tiles were made, moulded with courtly or religious figures in a good range of colours. Highly decorated lead-glazed flasks and jugs were produced in the 16th and 17th centuries. Another decorative technique which was very popular at this time was trailed slipware; this consists of pouring semi-liquid clay (slips) in quick bold patterns

A Staffordshire salt-glazed teapot, c. 1760. In a burst of creative and commercial energy the Staffordshire potters refined and developed new types of pottery that led to the creamwares and wide output of the Industrial Revolution.

A white stoneware tankard from Sieburg, Germany, with silver mounts probably made in England c. 1580. These finely crafted and durable wares were widely exported and the esteem in which they were held is evident from the quality of the silver mounts that were sometimes added.

under a transparent lead glaze. The technique produced energetic designs of great immediacy; slipware masterpieces include the Staffordshire chargers by Thomas Toft and his contemporaries, which have inspired the work of modern potters.

Stoneware production

Stoneware is fired to a higher temperature than earthenware and is hard, resonant and resistant to acids. It is also non-porous even when unglazed. One of the first methods of glazing stoneware was by throwing a handful of salt into the kiln where it vaporized and reacted with the surface of the vessels to form an extremely hard, tight-fitting glaze of great durability. Salt-glazed stoneware was made in Germany from the end of the 14th century and fine wares were made at Dreihausen and Siegburg during the 15th century. At Siegburg a fine white clay body was developed and medallions of crisply applied decoration were used. The same decorative techniques were used on the brown bodies of the Raeren pieces that were exported throughout northern Europe; these were sometimes fitted with fine silver mounts, which suggest that the wares were held in high regard. Alongside these fine wares there was a large production of flasks for wine and ale, which were decorated with a relief bearded face on the neck of the bottle. In England these were called "Bellarmines". Further forms of decoration were developed at Westerwald where blue and later manganese purple were used and at Kreussen where polychrome enamels were used to paint bands of classical or religious figures around the bodies of tankards.

The import of large quantities of German stoneware to England stimulated local production and by the end of the 17th century fine wares were being produced in London, Staffordshire and Nottingham. The Elers brothers' unglazed redwares, made in imitation of the Chinese Yixing pottery, were particularly finely potted and were much imitated throughout the 18th century.

Pottery and the Industrial Revolution

In Staffordshire during the 18th century there was a period of experimentation that, in conjunction with the birth of the Industrial Revolution, was to transform the production of pottery. The proximity of coal and a variety of local clays stimulated the production of many new types of earthenware and stoneware. Finely potted agate wares, made from mixed clays of different colours, lead-glazed teawares and superbly potted white salt glaze wares were made in profusion. Fanciful designs abounded, such as teapots in the form of animals and vegetables.

One of the greatest experimental potters was Thomas Whieldon, who made refinements to cream-coloured earthenware, 18th century England's most important contribution to the ceramic tradition. Josiah Wedgwood (who was in partnership with Whieldon from 1754–1759) together with others, developed printed forms of decoration and introduced his famous "basalts" and jasper wares which are still produced by the same firm today. Creamware and the whiter pearlware were made in vast quantities in Staffordshire and elsewhere in England, notably in Leeds, and exported widely throughout Europe.

Throughout the 19th century pottery was in competition with porcelain, now mass-produced. It was, however, still used, from the most elaborate wares such as the Minton "Majolica" extravaganzas to the hugely popular Staffordshire figures. After a period of comparative decline, pottery was to be revived towards the end of the 19th century with the work of the Art and Craft potters and, slightly later, the early studio potters.

INFLUENCES

1

2

*An altarpiece (**1**) by Raphael 1483–1520. The potters of the Renaissance strove to emulate the great painters of the period.*

*The import of large quantities of Chinese "kraak porselein" (**2**), c. 1600, led to its widespread imitation by the Delft potters of northern Europe.*

3

4

5

*A Meissen Augustus Rex vase (**4**), 1730–35, painted by A.F. von Lowenfinck. The secrets of porcelain manufacture discovered in Europe in the early 18th century stimulated tin-glaze potters to refine their art.*

*A shield in gilded bronze (**5**), Italy, 16th century, decorated with Apollo's chariot. Metalwork in all its forms was a constant influence on pottery – they competed with each other as a choice for tableware, particularly dishes and bowls. The decoration of this shield compares with the Istoriato style of Renaissance maiolica.*

*A bowl painted with a Coptic priest (**3**), Egypt, early 12th century. The Islamic pottery tradition bequeathed to Europe the lustre and white tin-glaze techniques.*

EARLY MAIOLICA

1

3

4

2

*A Faenza dish (**1**) dated 1497, painted with the arrival of Aeneas at Delos. The potters of Faenza refined the white tin glaze and developed a rich range of high-fired colours.*

*A Florentine oakleaf drug jar (**2**), c. 1440. The form of these jars was created in Italy but the decoration owes much to the Hispano-Moresque tradition.*

Spanish lustre wares were amongst the earliest fine European pottery. They were widely exported and held in very high regard.

*The Hispano-Moresque charger (**3**), c. 1469–79, is painted with the arms of Isabella of Castille and Ferdinand, King of Sicily.*

*The vase (**4**) is of the "Alhambra" type, c. 1465, and is painted with the arms of the Medici family.*

5

*The Florentine dish (**5**), second half of the 15th century, painted with the arms of the Guiducci family, and the Tuscan two-handled albarello (**6**), c. 1470, show the technical and artistic heights to which the craft of the potter rose during the later 15th century.*

*The earlier wares, such as this Orvieto ewer (**7**) of the early 15th century, relied on a simple palette and Gothic styles of decoration.*

7

6

ITALIAN RENAISSANCE

1

2

3

*An Urbino trefoil cistern (**1**), dated 1608, from the pottery of Francesco Patanazzi. It is painted with God confronting Adam and Eve within a border of "grotesque" decoration that was popular in the late Renaissance.*

*A lustred plate (**2**), 1525, perhaps painted in the workshop of Maestro Giorgio at Gubbio after Marcantonio's engravings of Roman heroes. The art of lustre decoration was largely confined to Deruta and Gubbio where decorated pieces from other workshops were brought to be enriched. Whether this dish was painted at Gubbio or brought from another centre is uncertain.*

*A late 16th century Montelupo wet drug jar painted with a roundel of St Anthony from the pharmacy of the convent of San Marco, Florence (**3**). Monastic pharmacies were important patrons of the maiolica workshops.*

*A broad-rimmed bowl (**4**) painted by Nicola da Urbino, Urbino c. 1520. This is an early work of one of the most important painters in the "Istoriato" or narrative style. The subject is an allegory in which a young woman, representing calumny, accuses an innocent prisoner.*

*A Deruta blue and gold lustre dish (**5**), c. 1520, showing Hercules lifting the giant Antaeus. The small town of Deruta was the largest producer of lustred wares.*

*A Florentine drug jar (**6**), late 15th century, painted with the arms of Pittigardi of Florence. The "bryony flower" decoration derives from the lustred Hispano-Moresque wares of Valencia.*

4

5

6

THE SPREAD OF MAIOLICA

1

*A Lyon Istoriato deep dish (**1**), c. 1580; painted with Moses receiving the tablets on Mount Sinai after the engraving by Solomon Barnard for Jean de Tourves' Lyon bible of 1554. Italian potters were recorded in Lyon as early as 1510 and after 1550 the number grew. This dish is very similar to the Urbino Istoriato wares and was certainly the work of an Italian immigrant.*

2

*A maiolica albarello or drug jar (**2**), c. 1540, probably Catalan. Spain already had an established maiolica tradition and the Italian influence is not always so overwhelming as in other countries. This example, however, follows the Italian style closely.*

*A Dutch dish (**3**), c. 1620. Maiolica in the Italian style was being produced in the Low Countries before 1500 but by the early 17th century distinctively North European styles had evolved.*

*This model of a fortified château from Nevers, France (**4**), dated 1689, is an example of the diversity of form and subject produced by Northern European potters of the period.*

3

4

DELFT

*A London delft royal portrait charger (**1**), c. 1665, painted with half-length portraits of Charles II and Catherine of Braganza.*

*The earliest dated (1602) English delft charger (**2**), is painted with the Tower of London and inscribed "The Rose is Red, The Leaves are Grene, God Save Elizabeth Our Queene!" Handsome chargers were made throughout the 17th and early 18th centuries and were purely for decoration.*

1

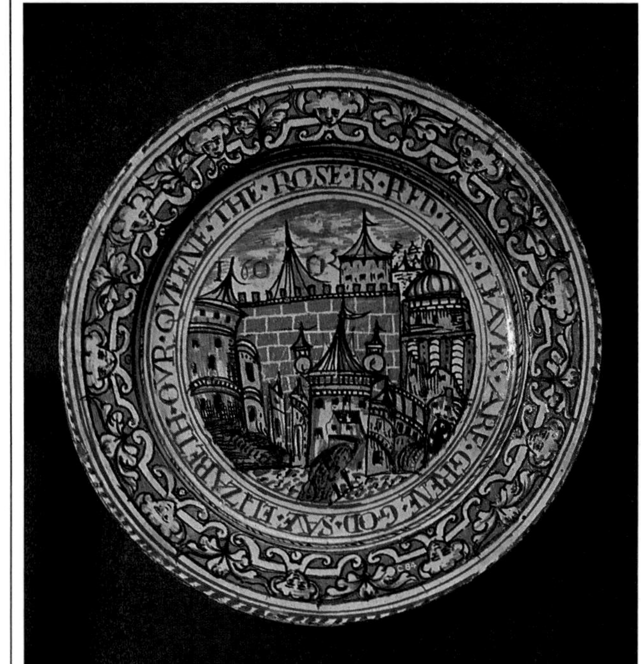

2

3

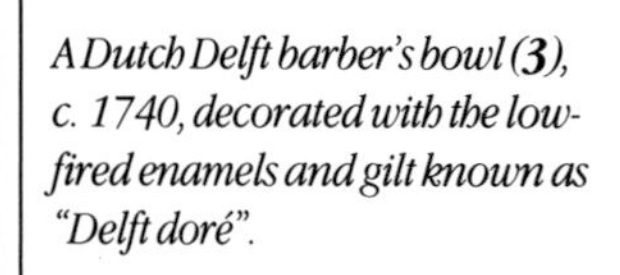

*A Dutch Delft barber's bowl (**3**), c. 1740, decorated with the low-fired enamels and gilt known as "Delft doré".*

*A Dutch sweetmeat dish (**4**), c. 1700, painted with a Chinese hunter in the "mixed technique" whereby the high-fired blue and white decoration is augmented with low-fired enamels and gilt. This dish would have formed part of a set of dishes of different shapes that fitted together to form a large centrepiece for the table.*

4

*A London delft royal equestrian charger (**5**), c. 1690. The subject is taken from a print by Cornelius van Dalen of Charles I entering Edinburgh, but perhaps in this instance James II is represented.*

*A Dutch "Delft-doré" dish (**6**), c. 1720, with the A.R. mark for Ary van Rijsselberg, painted with a subject commonly found on Chinese famille verte porcelain.*

5

6

7

*A Dutch Delft tulip vase (**7**), c. 1690, made in three sections that fit together to form a tall obelisk. The Dutch potters drew much of their decorative inspiration from the imported Chinese porcelains which were arriving in great quantity; the strongly architectural form is European in origin.*

*An English delft charger (**8**), Southwark, dated 1650, painted with the arms of the Vintners' Company. The fluted form derived from a metal original is found throughout Europe.*

8

FAIENCE

1

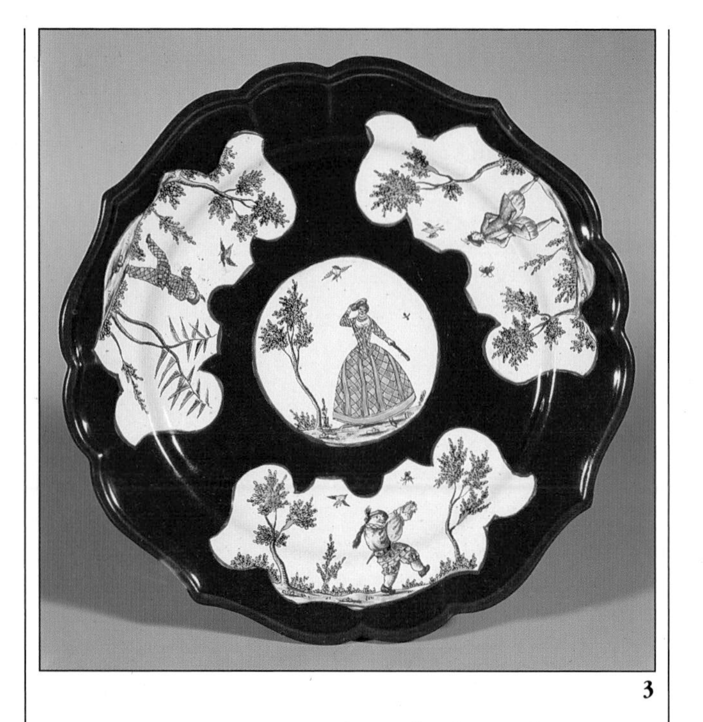

3

*A Strasbourg tureen in the form of a pigeon (**1**) after a model by J.W. Lanz, 1750–54. In France the partial royal monopoly on the production of porcelain stimulated the manufacture of fine tin-glazed wares.*

*Marseilles faience: a tureen (**2**), Veuve Perrin, and a plate (**4**), Fabrique de Robert, both c. 1765.*

*An Italian faience plate (**3**), Milan c. 1760, painted with figures from the Italian commedia dell'arte in low-fired enamels. Factories in Italy adopted low-fired enamels in the 18th century when the earlier maiolica traditions were in decline. As on porcelain the commedia dell'arte was an ever popular source of inspiration.*

2

4

*A German faience tureen from Fulda, (**5**), 1741–44, probably painted by Carl Heinrich von Lowenfinck. In Germany at this period only Meissen was producing porcelain, but both fine enamelled faience and a great deal of blue and white faience was made.*

5

6

*A Strasbourg dish (**6**), from the service made for the Elector Clemens Augustus of Cologne, 1750–51. The finest painting on faience equalled that on porcelain.*

LEAD-GLAZE AND STONEWARE

1

*A Nuremberg lead-glazed jug (**1**), c. 1550, from the workshop of Paul Preuning. The upper frieze depicts the adoration of the Magi and the lower shows the massacre of the innocents. This technique is often found on the stove tiles of the period.*

3

*A Fulham stoneware tankard (**3**) with relief decoration, inscribed "Drink to the pious memory of good Queen Anne, July 1729". Stonewares of this type continued to be made well into the 19th century.*

2

*A lead-glazed earthenware dish (**2**), c. 1550, by Bernard Palissy of France. These naturalistic wares had elements moulded from real animals and plants.*

4

5

*A pair of Staffordshire salt-glazed stoneware swans (**5**), c. 1750.*

7

*A Staffordshire slipware dish (**7**), showing a figure of Catherine of Braganza. By Ralph Simpson, late 17th century.*

6

*A Kreussen flask (**4**), early 17th century. The enamel decoration of this type of stoneware is closely related to that of contemporary German glass.*

*An Italian lead-glazed "sgraffito" dish (**6**) from Ferrara or Bologna, c. 1480–1510. At its best the "sgraffito" technique could rival contemporary maiolica.*

1

*A creamware cauliflower-moulded coffee pot (**1**) of Wedgwood/Whieldon type, c. 1760. The potteries in Staffordshire played an important role in the Industrial Revolution, and the 18th century was a period of considerable technical innovation. Creamware was widely exported and lent itself well to mass production.*

*A Whieldon creamware figure of a pug (**2**), c. 1760. Human and animal figures were popular, on both pottery and porcelain, to which the quality of the best creamware was comparable.*

2

3

*A Staffordshire salt-glaze pew group (**3**), c. 1740. Durable white salt-glaze was refined and much used in the middle of the 18th century, though it was later eclipsed by creamware.*

*Two teapots (**4**): an agateware teapot in the form of a pecten shell, c. 1745 and a Whieldon teapot in the form of an apple, c. 1755.*

5

*A Leeds creamware teapot (**5**), c. 1765, painted in black. The production of creamware spread to many centres in England, notably Leeds. Creamware lent itself well to painted and printed decoration.*

4

19TH CENTURY ENGLISH

*An Obadiah Sherratt group of "Polito's menagerie" (**1**), c. 1830. The Staffordshire potters produced popular figurative work throughout the 19th century, whose charm and vitality make up for what it lacks in technical execution. Amongst the most imaginative and entertaining of these potters was Obadiah Sherratt. His most elaborate works, such as this, with its characteristic stepped base, are folk art of great spirit.* **1**

*A Minton "majolica" jardinière and stand (**3**) designed by Albert Carrier de Belleuse, 1872. The firm of Mintons dominated English ceramics from about 1840 and many French artists, some from Sèvres, contributed to the high level of achievement. The glazed pottery known as majolica, was used largely for decorative works, though it was later used for domestic wares and architectural features.*

2

*A pair of Staffordshire ironstone vases (**2**), c. 1830. Ironstone was one of the new wares developed in Staffordshire and was patented by Charles Mason in 1813. It was alleged to contain slag of iron. Ironstone is thickly potted and very durable and was usually decorated in bold lively colours. It is still produced today.*

3

A Vincennes teapot, France, c. 1745–1750.

CHAPTER · SIX

EUROPEAN PORCELAIN

"God our creator has turned a gold-maker into a potter."

AN INSCRIPTION ABOVE THE DOOR OF J.F. BÖTTGER,
THE CREATOR OF TRUE PORCELAIN IN EUROPE

EUROPEAN PORCELAIN

It is difficult for us to appreciate the excitement and wonder in the courts of fifteenth century Europe caused by the arrival of Chinese porcelain. No comparable material was known of such hardness, translucency and brilliant whiteness. Marco Polo, who was the first European to witness the manufacture of porcelain, gave it the name "porcelaine", which originally meant a form of seashell. A gift of Chinese porcelain from the Sultan of Egypt was received by the Doge of Venice in 1461 and a further gift to Lorenzo de' Medici arrived in 1487.

The first successful attempt to imitate Chinese porcelain occurred between 1575 and 1587 in the Medici workshops. Medici porcelain, as it was known, was not true or hard-paste porcelain as had been perfected in China, but a close approximation called soft-paste porcelain. There are only about sixty known pieces of Medici porcelain recorded and all but two are decorated in blue and white; the forms and decoration are very varied and show the influence of Chinese and Middle Eastern wares. Most pieces betray their experimental origin with some imperfections. No further significant attempts to make porcelain took place in Europe for a hundred years.

Both hard and soft-paste porcelains were made within Europe, so it may be helpful at this point to describe them. "True" or hard-paste porcelain is formed from white china clay and fusible china stone. The china stone melts at temperatures of between 1300 and 1400°C and when cool holds together the finely dispersed particles of china clay. The glaze is also formed from china stone with the addition of a fluxing agent such as potash with lime or chalk to make it more fusible. Soft-paste porcelains generally substitute some form of glass for the china stone and are fired at a slightly lower temperature. The glaze is also of a glassy nature. Being partly of the same nature as the body, the glazes of both hard and soft-paste porcelain fuse with the body, and unlike earthenware glazes, do not form a completely distinct layer.

In 1673 Louis Poterat of Rouen was granted a patent for the manufacture of porcelain. Little can be ascribed to this manufactory with any certainty, but towards the end of the seventeenth century the Saint-Cloud factory, near Paris, began to produce fine soft-paste porcelains usually decorated with blue and white lambrequins. These French soft-paste porcelains are amongst the glories of European porcelain; beautiful glazes were developed which vary in tone from smoky ivory to brilliant white, and they have sympathetic warmth that was not matched in hard-paste porcelain. The factories of Chantilly and Mennecy (or Villeroy) were founded between 1725–1750 and drew inspiration from Oriental prototypes. French porcelain was to reach its peak later at Vincennes and Sèvres – but in the meantime, the reinvention of "true" porcelain in Germany eclipsed the French wares.

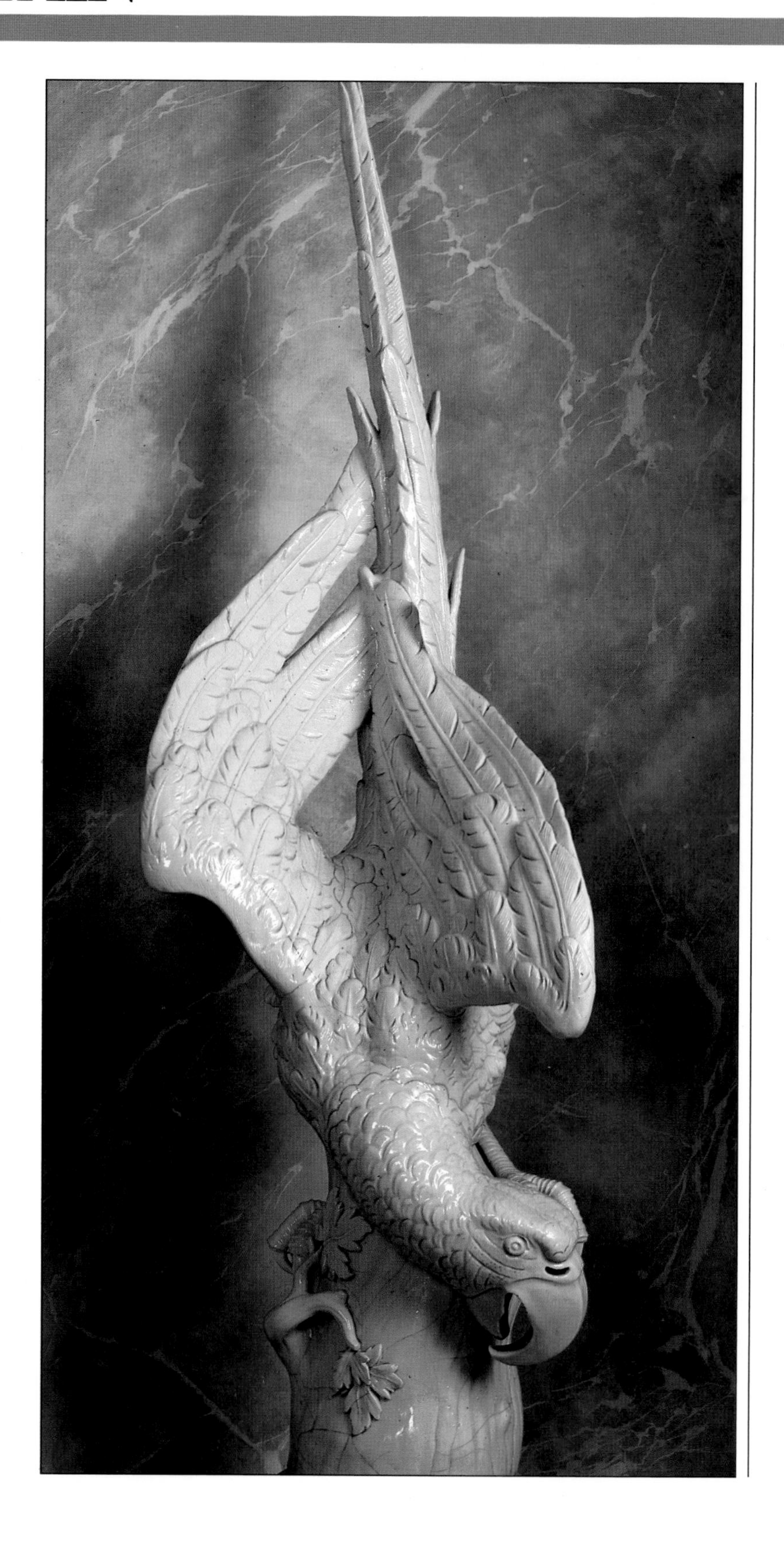

A Meissen centre dish from the "Plat de ménage" of Heinrich Graf von Brühl, c. 1737. The great baroque services of the 1730s were worthy vehicles for the talents of J.J. Kändler. Here the brilliant enamels and exotic chinoiserie combined to create a service then unparalleled in its splendour.

A Meissen white figure of a parrot, c. 1731. The great series of animals made for the Japanese Palace of Augustus the Strong was conceived on a scale that had hardly been attempted before, even in China. The technical problems of firing such a large mass of porcelain were huge and success was only partial. Meissen was fortunate in acquiring the services of J.J. Kändler, whose observations of the Royal menagerie at Dresden were brilliantly incorporated into the best of these figures.

Meissen porcelain

The Meissen porcelain factory, founded in 1710 in Saxony, was the first maker of true porcelain in Europe, and is still in production today. Until the Seven Years War in 1756, Meissen produced a great quantity of wares and figures of the highest artistic and technical qualities, and pioneered many new developments that were to have the greatest possible influence on European porcelain. For the first forty years of its production Meissen had no serious commercial rivals and its dominance was complete.

The principles of manufacturing hard-paste porcelain were "rediscovered" at Meissen through the experiments of a young alchemist, John Friedrich Böttger. His investigations of the properties of minerals and clays were done under the patronage of Augustus the Strong, the Elector of Saxony, who had a passion for collecting Oriental porcelain.

The first ceramic ware to be produced as a result of Böttger's experiments was the famous red stoneware in 1707. This had an extremely hard and fine-grained body ranging in colour from brick red to a very dark brown and was related to the red stonewares of Yixing in China that had been imported in the form of teapots. Böttger's red stoneware could be cut and polished on a lapidary's wheel and with the aid of the court silversmith, Irminger, most handsome forms were created. A black glaze that could be decorated with lacquer was discovered shortly afterwards. Further experiments led to the production of white kaolinic glazed porcelain of the Chinese type in 1709.

The Meissen factory was fortunate in gaining the services of Johann Höroldt in 1720, a talented painter who developed a full and brilliant range of enamel colours. The porcelain body, which previously had a slightly smoky hue, was improved in 1725 and led to a brilliant white porcelain, a perfect ground for the many skilled painters who were attracted to the manufactory.

A whole new repertory of designs and forms appeared in an explosion of creative energy. The Kakiemon porcelains of Japan were copied widely; chinoiserie decoration and harbour scenes were used on teawares. The patron of the factory, Augustus, inspired ever more ambitious creations and in 1731 a sculptor of genius was appointed, Johann Joachim Kändler. The great baroque porcelain sculptures of Meissen include the life-size animals made for the Japanese palace. Small figures, sometimes used for table decoration, were widely produced too, such as the "Italian Comedy" series upon which much of the factory's reputation and fame rests. The extensive dinner services and garnitures of vases also bear witness to the huge output and creativity of the factory.

The "arcanum" or secrets of porcelain production were jealously guarded, but runaway workmen succeeded in establishing two early factories. At Vienna, in 1719, the Du Paquier factory was founded. Some of the most delightful and original of European porcelains were made here in the baroque style, but never in large enough quantities to compete with Meissen. The Vezzi factory in Venice, founded in 1720, was an even smaller concern and the pleasing, if sometimes slightly primitive, wares are rare today.

Meissen's monopoly began to give way in the 1750s with the founding of seven important German factories, including Höchst, Frankenthal and Fürstenburg. Each produced hard-paste porcelain of excellent quality in distinctive and often admirable styles. The greatest masterpieces of this period are the figures modelled by Franz Anton Bustelli at Nymphenburg. These were in the new rococo style, and show a lightness of touch and delicacy of modelling, best illustrated in the white religious figures and brightly coloured "Italian Comedy" series.

Sèvres porcelain

The costs of producing the very finest porcelains in the 18th century were so great that major manufactories required patrons of enormous wealth. The most opulent court that Europe had ever seen was that of the Kings of France at Versailles and so it is not surprising that the French porcelain manufactories rose to dominance under royal patronage. The manufactory at the Royal Château of Vincennes is said to have started in 1738, but little is known of its products until the late 1740s. A soft-paste porcelain of exceptional beauty was developed there at great expense which, if it did not quite match the crisp perfection of Meissen, had a warmer, subtle quality which many people find more sympathetic.

White figures of classical inspiration displayed porcelain to great advantage and were widely made. The earliest wares were often simply formed but painted with exquisite delicacy in a subtle range of colours with floral sprays or Meissen style harbour scenes. The rococo style was now established and found great expression in the wares of the early 1750s. In 1756 the royal factory moved to Sèvres. Sèvres porcelain is usually dated and signed by the painter, and often the gilder, whereas at other manufactories the artists usually remained anonymous.

Madame de Pompadour was amongst the manufactory's greatest patrons and with such influence Sèvres could call upon the skills of the most talented artists of the day, amongst them Jean-Baptiste Oudry and Francois Boucher. The court goldsmith Duplessis and the sculptor Falconet were amongst the many talented artists and craftsmen attracted to Sèvres.

The development of ground colours, which had previously been left white, added a new dimension to Sèvres porcelain, and had a widespread influence across Europe. Notable amongst the ground colours were the deep mottled blue known as "Bleu Lapis", the sky blue "Bleu Celeste" and the delicate pink now known as "Rose Pompadour". The great refinement of the earlier wares gave way to a more grandiose style, suited to the palace settings for which they were destined. The quality of design and craftsmanship at Sèvres dominated European porcelain production for most of the rest of the century, with the neo-classical style eventually superseding rococo designs. The introduction of hard-paste porcelain in 1769, which was less costly to produce, helped to contain the ever-escalating expenses of the factory, but could not match the earlier soft-paste porcelain for beauty.

A Sèvres plate, 1823, showing work in progress in the painting and decorating studio.

A Worcester blue and white pierced basket, c. 1756. Worcester was one of the most commercially successful of the English porcelain factories. A great part of its production was blue and white wares, often painted with delightful chinoiserie scenes that, particularly in the factory's early years, rival the more elaborate coloured wares in beauty.

Italian porcelain

Elsewhere in Europe porcelain manufactories were set up during the 18th century and if they never achieved the stature of Meissen and Sèvres, much notable work was done and, on occasions, masterpieces produced. In Italy hard-paste porcelain of a greyish hue was made in the Venice factories and at Doccia where fine sculpture and finely painted wares were made. The greatest Italian porcelain is that of the soft-paste manufactory of Capodimonte at Naples, founded by Charles de Bourbon, the King of Naples and Sicily; this was a beautiful paste which came close, at its best, to that of Sèvres. Skilled painters and a modeller of genius, Giuseppe Gricci, combined to create works of great beauty that are now amongst the most expensive and sought-after porcelains of the 18th century. The factory moved with the King on his accession to the throne of Spain, where it became known as Buen Retiro, but it never regained its previous splendour. Fine neo-classical porcelain was made towards the end of the century in Naples.

English porcelain

Porcelain in England never enjoyed royal patronage on the scale of the European manufactories, but was brought about through the efforts of entrepreneurs. The earliest and most significant manufactory was Chelsea, created largely by the Huguenot silversmith Nicholas Sprimont. Although the earliest pieces, dating from 1745, are somewhat experimental they have a refreshing quality after their grander Continental counterparts. By the early 1750s a very fine paste had been developed and the quality of painting was high; the fable painting attributed to Jeffrey Hamet O'Neale is notable. The porcelain of this period can bear comparison with that produced in Europe. After about 1758 the factory embarked on some ambitious works in the Sèvres style, but regrettably without the Sèvres genius for design.

Some of the English figure modelling of the early 1750s was of a very high order. At Derby a series of chinoiserie figures, at Bow the works of the "muses modeller", and the wonderful white groups from the mysterious factory that produced the "Girl in a Swing" (believed to be related to Chelsea) show some anonymous modellers of great skill were inspired by having porcelain at their disposal. After these first few years most figures produced in England were copied from Meissen. Repeatedly one can see the inspiration of a factory's earlier years giving way to commercial pressures, and a lowering of artistic standards.

Much of the porcelain produced in England was for middle-class markets, and it was consequently less grandiose in scale and design than the European Court porcelain. But a large part of its considerable charm lies in its unassuming character, and it is still much admired for these qualities.

As the neo-classical influence began to permeate Europe towards the end of the 18th century porcelain became increasingly available and it lost some of its mystery and allure. There was, however, always a demand for works of high quality and from the late 18th and early 19th centuries works of the highest technical achievement were still made, often skilfully painted by miniaturists.

In the 19th century England pioneered the mass production of porcelain, and manufactories such as Worcester, Coalport and Spode exported their wares throughout the world. The quantities now produced were enormous and if much of it was rather routine the finest work still maintained a very high standard. The Victorian taste for elaborate decoration was reflected in The Great Exhibition of 1851 – where many of the pieces included must, by any standard, be regarded as masterpieces of inventive design.

INFLUENCES

1

*The interior of Ottobeuren Abbey, Bavaria (**1**), designed in 1737 by Johann Michael Fischer and built between 1744 and 1767. The spirit of rococo and baroque architecture inspired a theatrical extravagance in the applied arts; this dominated porcelain designs in Germany and the rest of Europe until the greater restraint of classical and French designs won favour in the latter part of the 18th century.*

2

Le thé à l'anglais chez le Prince de Conti *by M.B. Ollivier, 1712–84, (**2**). The fashion for tea drinking led to a demand for porcelains of great elegance in courtly circles.*

*A Japanese kakiemon dish (**3**), c. 1690, painted with the "hole-in-the-well design". Kakiemon wares were avidly collected in the early 18th century and many European factories copied them with considerable success. The jewel-like enamels and strong asymmetric designs were unlike anything hitherto seen in Europe.*

TEA AND TEAWARES

Tea became widely used in Europe after the middle of the 17th century; it was imported from China, along with the brown stoneware teapots known as Yixing ware. These imported wares determined the shapes of most early European teapots and red stonewares were soon made in Germany, Holland and England in imitation of the Chinese. Porcelain was especially suitable for teawares as it could withstand boiling water and was completely non-porous. The care lavished on the tea service was commensurate with the highly fashionable nature of tea drinking. As well as the teapot itself, sugar bowls, tea caddies, milk jugs, tea bowls and saucers were required. Coffee and chocolate drinking also demanded suitable wares so a fashionable household had to have a great range of porcelain. The frivolous social spirit of tea drinking was echoed in the rococo designs that swept through Europe in the 18th century, and no contrast could be greater than with the serious and austere Japanese tradition of tea drinking discussed in Chapter 3.

*An octagonal silver coffee pot (**4**), English, by Thomas Tearle, 1720. Silver competed with porcelain for tea and coffee services and tablewares and many of the early porcelain designers were themselves silversmiths.*

EARLY EXPERIMENTS

*A Böttger red stoneware pilgrim bottle (**1**), c. 1715. J.F. Böttger developed this highly fired stoneware in his attempt to produce true porcelain. The fine hard body could be cut and polished on a lapidary's wheel.*

*A Böttger stoneware teapot (**2**), c. 1715, after a design by J.J. Irminger. A black glaze was produced that could be decorated with lacquer. The form of the teapot derives from a silver original.*

*An opaque glass tumbler (**3**), English, perhaps London, c. 1760. Glass whitened with tin oxide gave a good imitation of porcelain and was the basis of many experiments in soft-paste porcelain.*

3

2

1

*A Medici porcelain ewer (**4**), c. 1575–1587, one of about seventy known examples of the earliest soft-paste porcelain made in Europe for Francesco de' Medici in Florence.*

*A Rouen mustard pot (**5**), late 17th century. The early production of soft-paste porcelains in France spread to a number of centres in the early 18th century.*

*A Böttger porcelain vase and cover (**6**), c. 1720. The earliest hard-paste or true porcelain produced in Europe.*

4

5

6

MEISSEN AND BAROQUE

*A tureen from the Swan Service (**1**), c. 1738, which was the greatest and most original of the Meissen services, created by Kändler and Eberlein for Heinrich Graf von Brühl, and comprising over 2000 pieces. In style it marks the transition from the baroque to the rococo.*

*A Böttger porcelain rectangular teapot (**2**), c. 1720, painted by Ignaz Preissler, one of the best of the many outside decorators ("hausmauler") who painted Meissen and other porcelains in their varied and individual styles.*

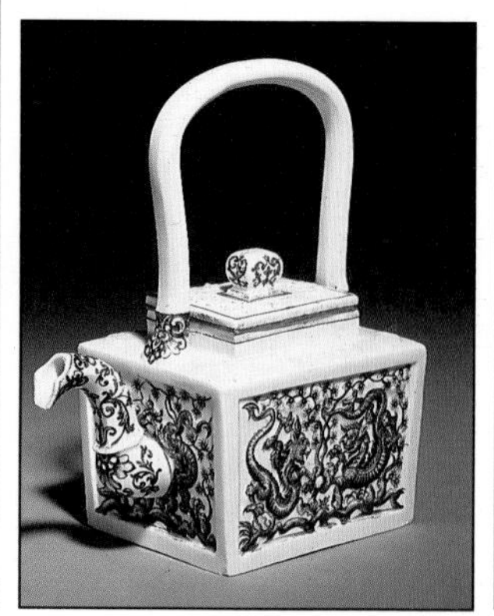

3

5

4

6

*An early example of an armorial teapot (**3**), c. 1730, bearing the arms of Christian VI of Denmark and chinoiserie decoration in the manner of J.G. Höroldt.*

*A figure of St John Nepomuk (**4**), c. 1731 by J.G. Kirchner, the predecessor of J.J. Kändler at Meissen. The figure shows the theatrical spirit of the baroque.*

*This dish from the "Christie-Miller" service (**5**), c. 1742, displays some of the finest and most elaborate painted decoration of the period. Harbour scenes were popular from the early 1720s.*

*An "Augustus Rex" vase (**6**), c. 1730-1732. This vase, made for Augustus the Strong, the Elector of Saxony, closely follows the shape of the Japanese Kakiemon original, but the painting is unmistakably Western in content and execution.*

THE SPREAD OF PORCELAIN

1

*The secrets of making porcelain eventually spread throughout Europe. Under the patronage of the King of Naples and Sicily the Capodimonte factory produced some of the finest soft-paste porcelains, such as these beakers painted by Caselli, c. 1747 (**1**). The Doccia beaker (**2**) c. 1742, bearing the arms of the last Medici, exhibits some of the best painting on Italian hard-paste porcelain. Chantilly (**3**) produced some of the most loved of the early French soft-paste porcelains; this "seau" (pail) c. 1745, typifies the prevailing fascination with the Orient. The short-lived Vezzi factory (**4**), 1720–27, was one of the first to acquire the secrets of true porcelain, because of the defection of Christoph Hunger from Meissen. In Germany, the 1750s saw the flowering of the art of figure modelling. This chinoiserie group from Höchst (**5**) dates from about 1753.*

2

3

4

5

VINCENNES AND SEVRES

*A Sèvres hard-paste tray (**1**), 1773, with the cipher of Paul Petrovitch. Hard-paste porcelain was used at Sèvres from 1769.*

1

*A Sèvres soft-paste biscuit group (**2**), c. 1765, of Leda and the Swan after a model by Etienne Maurice Falconet derived from a painting by François Boucher. Unglazed or biscuit porcelain was widely used to create figures and groups which were often used as table decorations.*

2

3

*A Vincennes tripod jug (**3**), c. 1745–50, painted in an early palette without the use of gilding.*

*A Vincennes écuelle, cover and stand (**5**), c. 1750. The superbly designed and executed early Vincennes wares are amongst the most beautiful of all European porcelains.*

4

A Vincennes "bleu lapis" ground cylindrical mug and cover (**4**), 1753, painted by Capelle.

6

A Sèvres plate from the Catherine the Great service (**6**), 1778. This service originally consisted of nearly 800 pieces and was the most expensive ever produced at Sèvres.

5

7

A Sèvres "cache pôt" (**7**), 1757, painted by Ledoux. The rose-pink was one of a number of distinctive ground colours developed at Sèvres in the 1750s.

ROCOCO

*A Chelsea teapot (**1**), gold anchor period, c. 1765. The elaborate design contrasts with the more austere works of the 1750s. Chelsea tried, with some success, to emulate the developments at Sèvres.*

*White figure of a girl beside a vase (**2**) Tournai, Belgium, c. 1765. The rococo style was well suited to the delicacy and scale of porcelain and is seen to particular effect when figures are left in the white.*

1

2

3

*A pair of Fürstenburg plaques (**3**), c. 1765, possibly painted by P. Weitsch. The frames display the characteristic asymmetry of rococo ornament.*

*One of a pair of Sèvres vases (**4**), c. 1760. The flamboyant designs at Sèvres could be strikingly original; here the use of chinoiserie decoration and two ground colours is particularly unusual.*

*A Frankenthal pot-pourri vase and cover (**5**), c. 1760. Extravagant rococo forms were popular in Germany in the 1760s.*

4

5

FIGURES

1

*The earliest Meissen figures (**1**) such as this farmer's wife, made in Böttger's early porcelain c. 1717, display considerable naïvety in contrast with the sophistication of the models by J.J. Kändler.*

3

*Amongst the masterpieces of English figure modelling, this Derby "dry edge" chinoiserie group (**3**) c. 1752–55 is from a set of the senses – it depicts the sense of smell. Most English figures were copied from Meissen and original work of this quality is rare. The sculptural nature of the group is enhanced by the lack of decoration.*

2

*The beautiful soft paste of Capodimonte porcelain (**2**) was a worthy vehicle for the talents of the great Italian modeller Giuseppe Gricci. "The Rabbit Catchers", c. 1750, typifies the aristocratic taste, so widespread in the 18th century, for peasant figures and idealized rustic life.*

4

6

*The Commedia dell'Arte inspired many of the finest porcelain figures. The Meissen Harlequin (**4**), c. 1740, is in the dramatic baroque style of J.J. Kändler, whose work was extensively copied at other factories.*

*The great modeller F.A. Bustelli at Nymphenburg brought the rococo style to its greatest perfection – his work displays exquisite fluency. This Scaramouche and Columbine (**6**) c. 1760, exemplify the vitality and brilliance of his figures.*

5

*These rare figures of Scaramouche and Ragonda from the Doccia factory (**5**), c. 1760, are strikingly effective with their dramatic use of black and gold.*

EARLY ENGLISH WARES

1

*A Chelsea botanical plate (**1**), 1752–55. This type of decoration has become known as "Sir Hans Sloane" botanical after the patron of the Chelsea Physic Garden and was often based on the work of Philip Miller and Georg Ehret. These delightful wares are notable for their botanical accuracy and the frequent choice of vegetables and humble country flowers, which can be a refreshing change from some of the more grandiose designs of their Continental counterparts.*

3

*A Chelsea crawfish salt (**3**) of the Triangle period, 1745–49, painted in the studio of William Duesbury. Nicholas Sprimont, the manager of the Chelsea factory, was a distinguished Huguenot silversmith and this salt was a close copy of the silver originals made for Frederick Prince of Wales. It is still in the Royal collection. The naturalism of these salts was a feature of the early rococo decoration of the period.*

2

*A Worcester bowl (**2**), c. 1752–4, painted in a palette resembling that of the Chinese famille verte. Although the Chinese influence is evident, the treatment is peculiarly European. In many factories in England and Europe the products of the first few years achieved a level of beauty that was rarely recaptured later, when the early inspiration had worn off and commercial pressures dictated the nature of the products.*

4

*The Chelsea teapot (**4**) of the Raised Anchor period, 1749–52, is decorated with a pattern known as "the lady in a pavilion" after a Japanese Kakiemon original. The Kakiemon designs were enormously popular in the early years of the Chelsea factory and their use of a few brilliant enamels and areas of undecorated porcelain was perfect foil for the beautiful Chelsea porcelain body of the period.*

5

*A Chelsea bust of a laughing girl (**5**) of the Triangle period. 1745–49, showing a fluency that was lost in much of the later work.*

*A Bow cylindrical mug (**7**), c. 1750. Blue and white porcelain did not require the further firings of enamelled wares and was cheaper to produce, so it was used for much domestic ware. It could sometimes achieve the level of artistry shown in this example.*

7

6

*A finely decorated Bow plate (**6**), c. 1760, painted by Thomas Craft. The vigorous painting is a loose interpretation of the Kakiemon style and includes a pink enamel not found on the Japanese wares. Painting of this quality is rare on Bow porcelain.*

NEO-CLASSICISM

1

2

3

*A Paris vase (**1**), c. 1785, by Fabrique de Locré.*

*A Sèvres octagonal plate (**2**), 1784, from the Arabesque Masson service. This service is said to have been commissioned by Louis XVI for Marie Antoinette. It introduced new shapes and neo-classical decoration and was designed by Louis le Masson after Roman and Pompeiian themes.*

*A Sèvres biscuit figure of "Le Chancelier Michel L'Hôpital" (**3**), c. 1784, after the sculpture by Goix from the series "Des Grands Hommes". The severe posture contrasts with the earlier rococo biscuit figures.*

4

*A Sèvres "Vase Bachelier Rectifié" (**4**), 1779, painted by Antoine Caton, the reverse by Charles Buteux. Sèvres pioneered the neo-classical style in its grandest form with vases intended for the royal palaces and as diplomatic gifts.*

*A soup tureen and cover from the Erolanese service (**5**), Naples, c. 1781. This service was ordered by Ferdinand IV, King of the Two Sicilies, and delivered to his father Carlos III of Spain in 1782. The inspiration for the service came from the collection of bronzes found at Herculaneum.*

5

1

2

3

*A Chamberlain's Worcester vase and cover (**1**), c. 1815, painted in the manner of Humphrey Chamberlain with the words "Wolsey, received at ye Abbey of Leicester". The wealth of the English aristocracy and the burgeoning middle classes created a market for the growing number of factories producing high quality wares.*

*Sèvres yellow-ground vases, (**2**) and (**3**), c. 1809, painted by Georget with a portrait of Laura, Comtesse Régnault Saint-Jean d'Angely as a muse. Reputedly a gift from Napoleon.*

4

*A massive Sèvres vase "Etrusque à rouleaux" (**4**), c. 1813, by Percier and painted by Beranger, showing the arrival of works of art from Italy at the Musée du Louvre in 1796. The grandeur of the French porcelain of this period reflects the spirit of the Empire at its height.*

5

*This Derby tureen cover and stand (**5**) was painted c. 1800 by William Pegg, the Quaker, one of the greatest of botanical painters on porcelain. The lavish and life-like decoration is enhanced by the quality of the porcelain body and glaze. This botanically accurate style of flower decoration on porcelain represents a peculiarly English tradition that reached its highest expression in Chelsea and Derby wares of the 18th century.*

19TH CENTURY PORCELAIN

1

*A Berlin royal presentation vase (**1**), 1861–71, with a portrait of Kaiser Wilhelm I probably painted by H. Looschen. The particular form of this vase was first produced in 1831 and continued in use for much of the 19th century.*

*A Sèvres plate from the "Service de la Production de Nature" (**2**), c. 1846, painted by J. Lejour.*

*A pair of Coalport vases (**3**), c. 1870–75, painted by John Randall. These are painted in Randall's later style in which birds were depicted in their natural environment rather than in the more formal Sèvres style.*

3

2

*A Minton "pâte-sur-pâte" vase (**4**), c. 1903, painted by L. Solon. The "pâte-sur-pâte" technique was developed at Sèvres and consisted of carefully built up layers of white slip which when fired became translucent in varying degrees according to the thickness of the layers. This vase shows the Peloponnesian King Lycurgus enthroned.*

*A Berlin vase (**5**), c. 1840–50 of "Französische" form. In the 19th century technical standards and craftsmanship reached a level that has not since been equalled.*

4

5

Baluster jar by William de Morgan, c. 1890.

CHAPTER · SEVEN

ARTS AND CRAFTS TO ART DECO

"I do not believe in the world being saved by any system – I only assert the necessity of attacking systems grown corrupt, and no longer leading anywhither."

WILLIAM MORRIS, 1883

ARTS AND CRAFTS TO ART DECO

The Great Exhibition of 1851, the first of the great international industrial exhibitions, was held in London and promoted by Henry Cole and the Prince Consort. The exhibition was a vast show-piece, intended to celebrate the industrial achievements of Britain, and indeed the technical mastery of much that was displayed was very impressive. But disquiet arose in intellectual and artistic circles as to the direction of the decorative arts. In the eyes of some leading thinkers, the prevailing high Victorian styles had become over-ornate and corrupt. Many attempts were made to evolve a new artistic language for ceramics, with the result that from this period and throughout the 20th century a very rich and varied body of work has been produced, absorbing a bewildering array of influences and ideas which continue to develop and interact.

One of a pair of tulip vases designed by William Burges for Cardiff Castle, 1874. William Burges was one of the more original of the English Gothic Revival architects, whose work for the Marquis of Bute at Cardiff Castle included furniture and ceramics. The Gothic Revival started in the mid 18th century and continued through the 19th century; its work was much admired and promoted by Ruskin, who was one of the spiritual fathers of both the Arts and Crafts and Pre-Raphaelite movements.

The Arts and Crafts movement

The Arts and Crafts movement in Britain derived its inspiration from the work of Ruskin and earlier thinkers, who felt that a soulless machine could not produce anything of artistic worth. This rejection of industrial production led to the revival of traditional craft pottery, inspired by a nostalgic view of the medieval craftsmen and guilds. There was also a desire to break down the divisions that had arisen between the artist and the craftsman. It was the business acumen of William Morris that actually put these ideas into practice. William Morris was a Utopian socialist whose ideas on social reform went hand in hand with a genius for design. His work had a widespread influence in Europe and the United States: the Wiener Werkstätte of Vienna, the rise of Art pottery in America and even the Bauhaus of Germany were all directly influenced by his ideas.

The potter most closely connected with William Morris was William de Morgan, whose main interest was in the design and decoration of wares rather than their form. Indeed, during the 1870s, others made and decorated simple shapes to his designs. He developed a range of lustre colours after Islamic and Hispano-Moresque earthenwares and an attractive palette of blues, greens and reds that he called Persian colours. Along with other workers in the Arts and Crafts and Pre-Raphaelite movements, de Morgan sought to break away from the classical European tradition and found inspiration in medieval Gothic art and the art of the Near East. Heraldic beasts, ships and floral designs with an Iznik flavour are the dominant motifs on his tiles and wares.

Doulton & Company was a large manufacturer of utilitarian salt glaze stonewares ranging from flasks and ginger beer bottles to drainpipes for the sewers of London. In 1871, in close collaboration with the Lambeth School of Art, an "Art Studio" was set up, with students from the school decorating the wares made by the factory potters. Notable amongst the many artists were Hannah Barlow, George Tinworth and Frank Butler. The good range of colours and varied decorative techniques developed led to an immediate and continuing success for the venture and by the end of the century over 300 people were employed in the Art Studio.

In 1873 the four Martin brothers set up a pottery in Fulham, moving later to Southall in Middlesex. They were unique at that time in that they only made "Art" pottery. Their stoneware was influenced by Doulton; the main production was in thrown wares with moulded and incised decoration, but they are best known for their comically grotesque face jugs and animals, particularly the wonderful grinning birds which are often in the form of tobacco jars.

Many Art potteries flourished towards the end of the

A William de Morgan lustre dish, c. 1885. De Morgan worked closely with William Morris in creating a new vocabulary of design; he sought inspiration from neo-classical sources, gothic design and particularly from Islamic and Hispano-Moresque lustre pottery. The revival of lustre techniques came after many centuries of neglect. The effect derives from firing metal oxides in a reducing atmosphere. De Morgan's interest in lustres came as a result of his experiments with the decoration of glass.

century and continued into the 20th century. The designs of Christopher Dresser, first at Minton's and Wedgwood, then at Linthorpe and Ault, played an important part in a widening repertory of influences. Notable also were the sgraffito wares of the Della Robbia factory, the high-fired glazes of the Ruskin pottery, and the lustres of Bernard Moore and the artists of the Pilkington Tile & Pottery Company.

American Art pottery

The important American Art pottery movement of the 1880s and 1890s was strongly influenced by the Arts and Crafts movement. But it also drew considerable influence from Japan and the early French artist-potters such as Ernest Chaplet, who had brought the rich Chinese flambé glazes to a high state of perfection. Cincinnati was the most important centre for Art pottery. In 1871 the ceramic chemist Karl Langenbeck and Maria Longworth Nichols began experimenting with ceramic colours and produced some handsome pottery with a group of amateur lady decorators. One of these, Mary Louise McLaughlin, developed a technique of painting in coloured slips that became known as Cincinnati faience.

The Rookwood pottery, established in 1880 by Maria Nichols, concentrated initially on her own work and some commercial lines. As it expanded a wide range of techniques and glazes were experimented with and developed, including the crystalline tiger's eye and Goldstone glazes. A technique of spraying atomized glazes, which allowed for subtle blending and changes of colours, was widely used, as well as much figurative and floral painting.

Fine high-fired Oriental crackled glazes were used at the important Chelsea pottery near Boston of the Robertson family. At the Lonhuda pottery, Ohio, American-Indian forms were introduced. The Grueby-Faience Company, established in 1894, initially produced wares in historical styles, but developed strong and austere shapes influenced by the work of the French potter Delacherche; these were based on natural forms that were moulded onto thrown vases. The matt glazes had characteristic mottled and veined effects. William Grueby adopted the social ideals of the Arts & Crafts movement and intended his business to be a "happy merger of mercantile principles and the high ideals of art". Adelaide Alsop Robineau was extremely influential as the publisher of the "Keramic Studio" and the maker of laborious and remarkable porcelains with incised and carved decoration and a range of fine glazes.

Industrial Design

A further important strand in the history of modern ceramics may loosely be called "Industrial Design", and is inextricably bound up with the Arts & Crafts movement. In the search for new sources of inspiration the Middle East and peasant pottery had been fertile areas, but it was above all Japan that brought a fresh decorative tradition untainted by European classicism. The Japanese influence can be seen in England in the production of the Royal Worcester Porcelain Company but it was perhaps more significant in France.

There designers like Felix Bracquemond, who worked at both Sèvres and Limoges, established the style that would later develop into "Art Nouveau". Towards the end of the 19th century in Copenhagen, the ware produced by both the Royal Porcelain factory and Bing & Grondhal gave a peculiarly northern flavour to the prevailing style; the Japanese influence, whilst rarely overt, can be seen in the asymmetry and use of undecorated space. A modern sense of abstraction entered the figure modelling.

A figure of Daphne, c. 1925, Berlin porcelain by Paul Scheurich. The New York born German modeller Paul Scheurich worked at Meissen, Berlin, Nymphenburg and later at the Rosenthal studio, and was one of a group of modellers who revived the art of figure modelling in a modern idiom. His graceful white figures are amongst the most successful of the period.

The Art Nouveau style was widespread across Europe and some of its finest expressions are found in the eggshell porcelains of the Rozenburg pottery at The Hague, Rorstrand in Sweden, Zsolnay in Hungary and at Sèvres. As early as 1901 the remarkable designs for dinner services by Peter Behrens at Darmstadt foreshadowed the more severe style of the 1920s and 1930s.

The large German porcelain factories sought to introduce modern designs and were particularly successful in creating some fine figure modelling, after a hundred years of simply reproducing the great works of the 18th century. The work of Paul Scheurich at Meissen, Nymphenburg and Berlin, as well as Ernst Barlach and Adolf Amberg, constituted a minor renaissance in this particularly German art.

The Bauhaus

The Bauhaus school of architecture and applied arts, established by Walter Gropius at Weimar in 1919, was to have an overwhelming influence on industrial design throughout Europe and America. The ceramic workshops were established at Dornburg, 25 miles from Weimar, in an area with an existing pottery tradition. In the hard post-war conditions the ceramic workshops initially attempted to revive the local traditions of form and craftsmanship, somewhat in the spirit of William Morris, under the crafts master Gerhard Marcks. Decoration was considered secondary to form. The Bauhaus respect for material was exemplified in Otto Lindig's well-proportioned works for practical use, designed to be suitable for industrial production. Simple geometric mould-made forms were developed, notably by Theodor Bogler, which would be combined in various ways to form useful wares; the intention was to simplify the process of industrialization. The workshop itself, in the absence of willing industrial manufacturers, produced its wares in large series in an attempt to raise much needed income. The first Bauhaus products to be produced industrially were Bogler's kitchen storage jars in 1923. Eventually, under the influence of Walter Gropius, the ideals of craftsmanship gave way to the creation of prototypes for machine production. When the Bauhaus moved to Dessau in 1925, the ceramic workshop was not transferred but continued at Dornburg. Otto Lindig eventually took over the direction and continued to preach with considerable influence his ideas of the superiority of form over decoration.

The Bauhaus idea that function should determine form was to have a significant influence throughout the West; when the Bauhaus was forced to close under pressure from the Nazis in 1933 some of the designers emigrated to America and other parts of Europe. Their ideas were widely disseminated and their influence continued long after World War II.

An Artěl coffee service, 1911. The Artěl craft workshops were established in 1908 in Prague along the lines of the Wiener Werkstätte and produced work in various mediums. The advanced design shown here is in the style known as Czech Cubism and foreshadows much of the work of European ceramicists in the 1920s.

In Scandinavia, notably Finland, functionalism produced classic designs of great purity and practicality which lent themselves well to industrial production. The Arabia factory of Finland has been particularly important in fostering modern designs by employing artist potters and allowing them to work in their own style whilst using the factory's facilities.

Other European developments

Cubist painters influenced the work of designers such as Michael Powolny of the Wiener Keramik and the emerging Art Deco style. Art Deco spread from France to most of Europe in the 1920s. The work of Clarice Cliff in England and the Doccia factory in Italy, are typical of the bold dynamism of that new style.

In France particularly, artists have frequently turned their hand to pottery, usually in collaboration with potters. Gauguin modelled wares that were fired and glazed by Ernest Chaplet and André Methey made pieces that were decorated by a roll-call of contemporary painters such as Rouault, Bonnard, Vuillard, Van Dongen, Derain and Vlaminck. The artist René Buthaud came to specialize in decorating ceramics. Perhaps the most fruitful collaboration was that of Pablo Picasso with the Madoura pottery at Vallauris from 1947. He worked on tin glazed and slip-painted dishes, then on novel vessels made to his design and also on figures of birds. In England Roger Fry, having learnt the basics of pottery from a country flowerpot maker, founded the Omega workshops in 1913 with Duncan Grant, Vanessa Bell and Wyndham-Lewis, to work in various mediums. The decoration of the wares was bright and exciting.

INFLUENCES

1

*A teapot (**1**) by Marianne Brandt, c. 1924. The Bauhaus developed a radical approach to industrial design. The geometrical form of this teapot is very similar to the work of Theodor Bogler of the Bauhaus ceramic workshop and was designed with a view to industrial production.*

2

3

*A Hispano-Moresque armorial charger (**2**), c. 1470. The unaffected strength of the Gothic style appealed to those potters who were trying to escape from the conformity of Victorian design. The lustre technique was revived after being neglected for 300 years.*

*Earth Mother (**3**) by Edward Burne-Jones, 1882. Burne-Jones worked with William Morris in the design of furniture and textiles; the pre-Raphaelite and Arts and Crafts Movement were closely related.*

4

*An Iznik dish (**4**), Turkey, early 17th century and a Japanese woodblock print (**5**) by Imao Keinen, c. 1892. Inspiration was sought from foreign cultures untainted by western classicism; the brilliant colours and designs of Ottoman Turkey and Japan were particularly influential.*

5

6

*The entrance of 30 Rockefeller Plaza (**6**). The sculptor Lee Lawrie collaborated with the ceramicist Leon V. Solon to create this triptych of* Wisdom, Sound *and* Light.

ARTS AND CRAFTS

*A Minton porcelain vase (**1**), c. 1880, designed by Christopher Dresser and made at the Minton's art pottery studio. Dresser designed in many mediums, drawing on a wide variety of traditions.*

*A Martinware tobacco jar and cover (**2**), c. 1900, in the form of a grotesque bird. The Martin Brothers were amongst the pioneers of the smaller art pottery manufacturers.*

1

2

3

*A Royal Lancastrian lustre vase ware, (**3**), c. 1910, painted by Gordon Forsyth. The Pilkington factory attracted a group of talented artists who worked with glaze effects and a particularly fine range of lustres.*

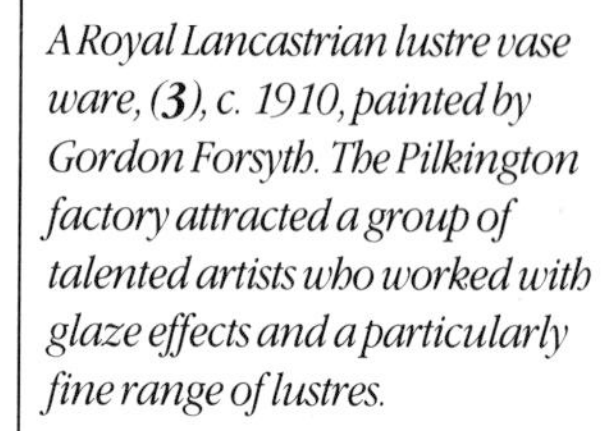

4

*A William de Morgan baluster jar (**4**), c. 1890, painted in "Persian" colours with a design of Iznik derivation. De Morgan worked closely with William Morris and was the most important of the ceramicists of the Arts and Crafts Movement.*

*A Linthorpe pottery vessel designed by Christopher Dresser in the form of a sea urchin (**5**), c. 1880. The stirrup handle is derived from pre-Columbian pottery.*

*A Doulton "Chang" ware vase by C. Noke (**6**), c. 1925. Doulton started as a manufacturer of utilitarian stoneware and then founded what was to become one of the most technically innovative art potteries.*

*A Doulton "faience" vase (**7**), c. 1880, decorated by Florence Lewis who specialized in painting flowers, foliage and birds.*

5

6

7

1

2

3

*A bowl by Frederick Hurten Rhead (**1**), c. 1915. Rhead was born in England where he worked for a number of years before emigrating to America. He set up his own pottery using local clays at Santa Barbara, California, working mostly in the sgraffito technique.*

*A green-glazed vase by Fritz Albert for the Gates Pottery (**2**), c. 1910, with a matt green glaze of a type pioneered by the Grueby faience company of Boston.*

*A vase by Jacques Sicard for S.A. Weller (**3**). Sicard studied the technique of iridescent glazes in France under Clement Massier. His "Sicardo" ware was made by the Weller pottery between 1901 and 1907.*

*A lidded vase called "Fox and Grapes" by Adelaide Robineau, c. 1920 (**4**). Robineau was an energetic experimenter with glaze techniques and decoration, as well as a potter of great originality. With her husband she founded the Keramic Studio which became very influential in the dissemination of new ceramic ideas.*

*A charger decorated by John Bennett (**5**), 1878. Bennett trained at Doulton's Lambeth faience pottery before moving to America in 1876. He painted flowers and foliage sketched from natural specimens on biscuit ware imported from England and later on cream-coloured earthenware made at his own workshop.*

4

5

6

*A vase designed by Ruth Erikson (**6**) for the Grueby faience company. Erikson was inspired by the French potter Auguste Delacherche; her natural forms and matt glazes were widely copied by other American art potteries. Here the decoration was modelled onto a thrown vase.*

ART NOUVEAU

*A stoneware vase (**1**) by Henri van de Velde, c. 1905. The Belgian-born designer was a key figure of the Art Nouveau and Modern movements. He helped to reform schools of applied art at Weimar where his pupils included Otto Lindig, who later worked at the Bauhaus.*

1

*A Sèvres "Vase de Montfort Orné" (**2**), in "porcelaine nouvelle" of 1908. Art Nouveau was at its height at the time of the Exposition Universelle in Paris, 1900, and continued to be fashionable through the next decade.*

2

3

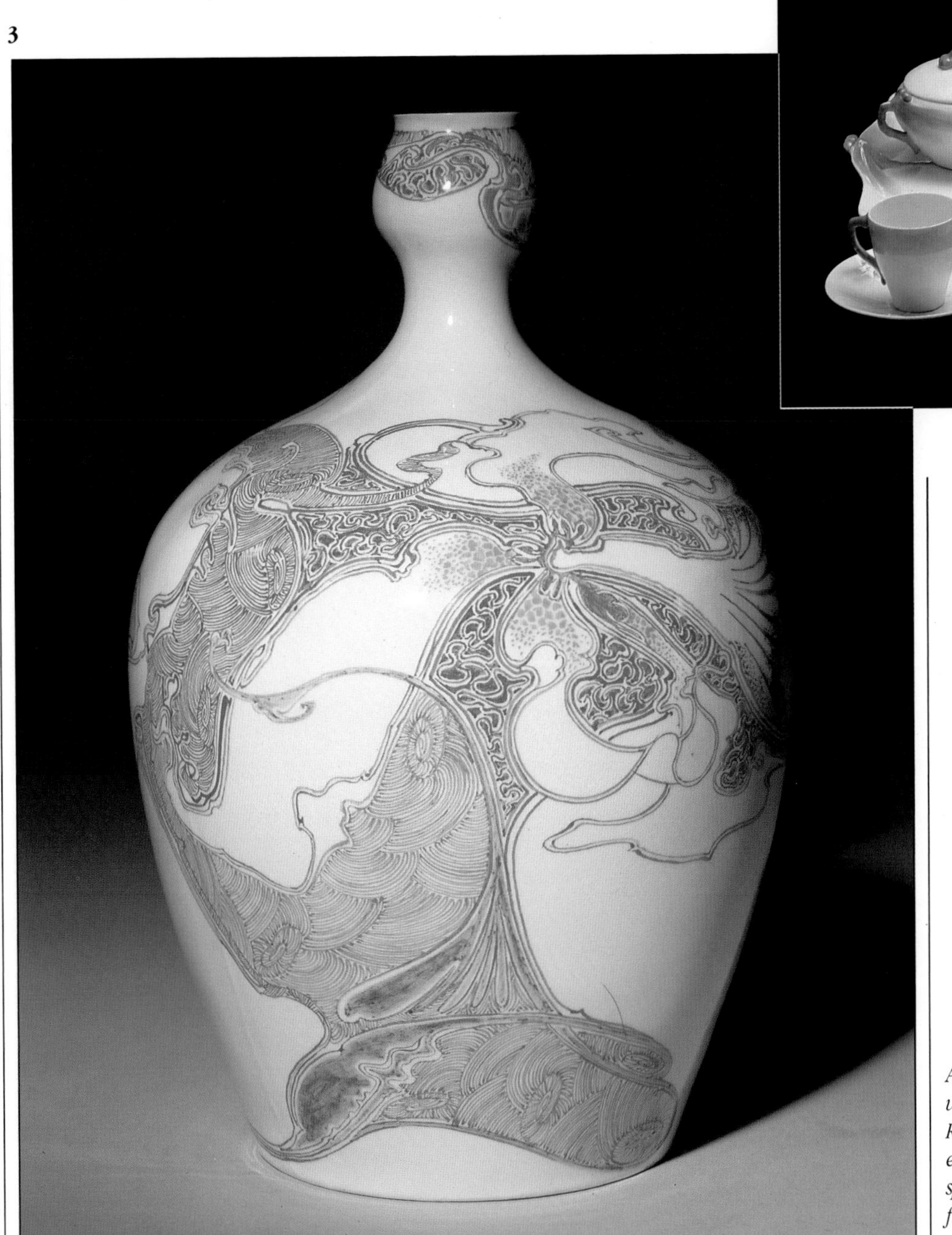

4

*A coffee service (**4**) by Alf Wallander, c. 1900. The Swedish-born artist trained in Stockholm and became a designer and later art director of the Rörstrand factory.*

*A Rozenburg eggshell porcelain vase (**3**), c. 1900, designed by R. Sterken. The Rozenburg pottery established at The Hague specialized in the production of finely potted wares in this distinctive and very fluid style.*

INDUSTRIAL DESIGN

1

A tea service (**1**) designed by Margarete Marks in 1930 and made in 1938 by Greta Pottery, Ridgway factory UK. Simple modular elements are combined in different forms.

*An earthenware bowl (**2**) designed by Keith Murray for Wedgwood & Sons, UK, c. 1935. Keith Murray, an architect, created a range of designs characterized by the decoration of grooves and stepped surfaces on simple forms in black basalt or in earthenware with matt or semi-matt glazes. He later designed the new Wedgwood factory itself.*

2

*A double handled punch bowl (**3**) by Theodor Bogler, Germany, c. 1925. The simple form relies solely on the contrasting bands of glaze for decoration.*

3

*A cup and saucer (**4**), c. 1923, by Otto Lindig, the technical director of the Bauhaus pottery. The simple form and decoration limited to a muted glaze effect typify the austere ideals of Bauhaus pottery.*

*A teapot (**5**), c. 1923, by Theodor Bogler. An aim of the Bauhaus workshop was to design works formed from simple elements which could be combined in different ways and so lend themselves to industrial mass production.*

*A stoneware tankard (**6**), c. 1906 by Peter Behrens, the German architect. It is in the early Art Nouveau style influenced by Henri van de Velde and Charles Mackintosh.*

4

5

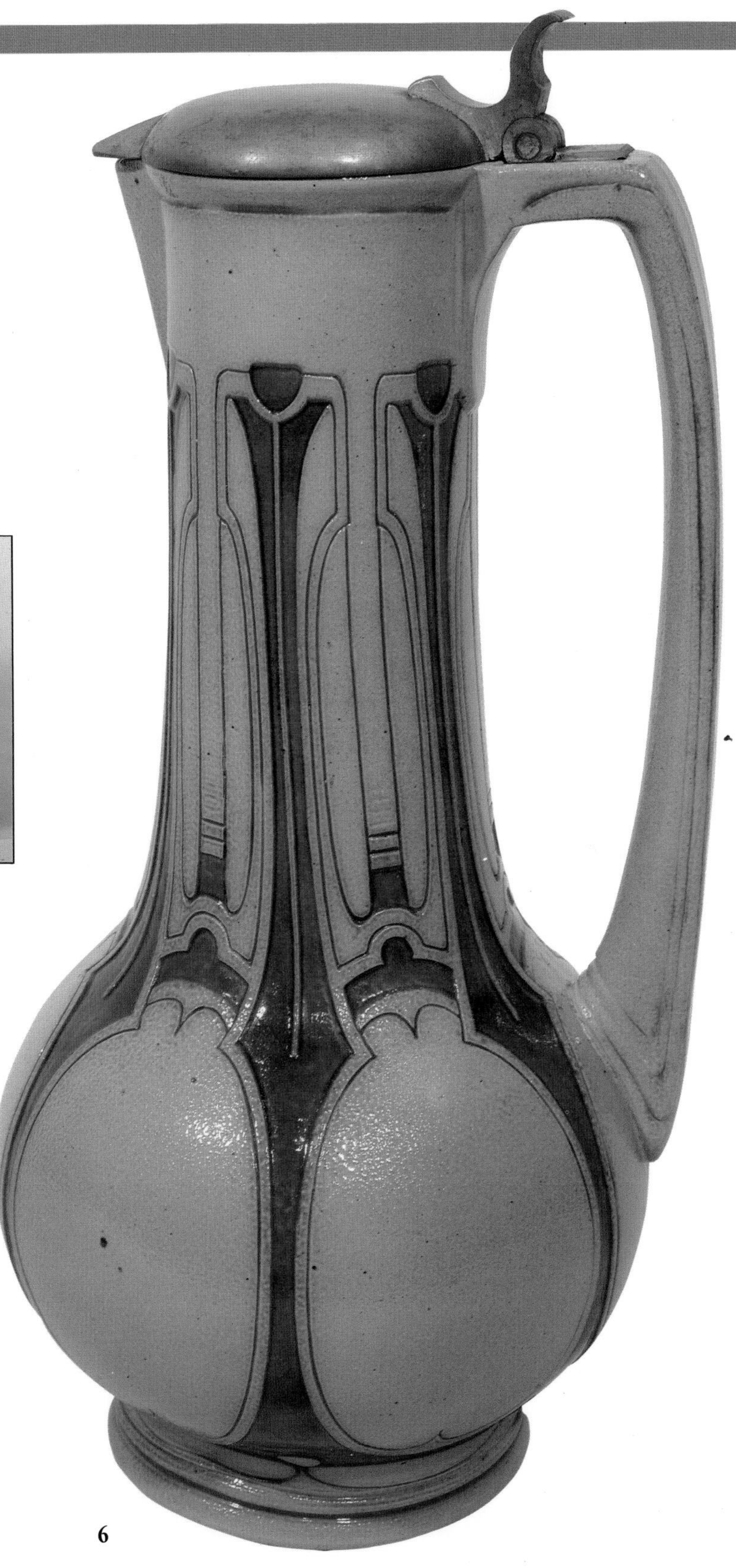

6

ART DECO

1

*"Omaggio Agli Snob" (**2**), a Richard-Ginori porcelain box and cover from Italy.*

2

3

*In the 1920s the new style swept through Europe, culminating in the Paris Exhibition of 1925. The Soviet propaganda plate (**1**) of 1921 is after a design by Alexandra Shchekotikhina-Pototskaya and is much influenced by contemporary art.*

*Four pieces (**3**) by Keramis and Boch La Louvière from Belgium with typical bold designs and bright colours.*

4

*Wiener Keramik figure (**4**), c. 1910, designed by Michael Powolny, an early pioneer of Modernism at Vienna.*

5

6

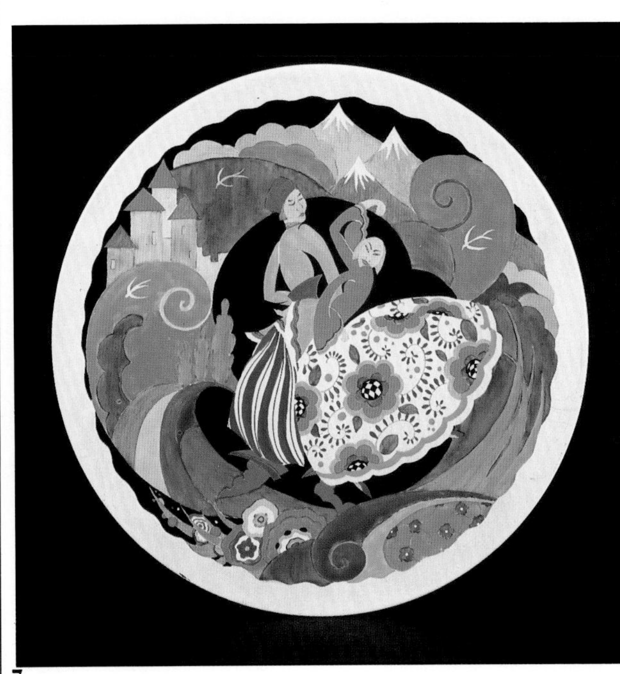

7

*Works by René Buthaud (**5**) of France, Viktor Schrekengost (**6**) of America and Clarice Cliff (**7**) of England. A rich diversity and strong internationalism are characteristic of this optimistic and forward-looking period.*

ARTISTS AS POTTERS

1

2

3

*A saucer by Wassily Kandinsky (**1**). Kandinsky brought his colourful abstract designs to ceramics. He said "the more abstract form is, the more clear and direct is its appeal". As with many artists working with ceramics, his work was more concerned with decoration than with form.*

*A stoneware dish in the form of a nude stepping down to bathe (**2**), 1887-88, by Paul Gaugin. Gaugin collaborated with the potter Ernest Chaplet between 1886 and 1889, initially decorating vases and mugs from Chaplet's normal output and later modelling the dark stoneware body used by the workshop into freely formed wares.*

Le Lapin, *a painted and glazed ceramic relief by Fernand Léger 1952 (**3**). Léger produced a number of impressive decorative designs for ceramics in his later years.*

Decore Feuilles by Pablo Picasso (**4**). Picasso, of all the artists of the 20th century, made the most significant contribution to ceramic art. In 1947 he began to work with Suzanne and Georges Ramié at the Madoura pottery in Vallauris and was at first concerned mostly with decoration and subsequently with forms made to his own design. His later output was varied and included much sculptural work.

4

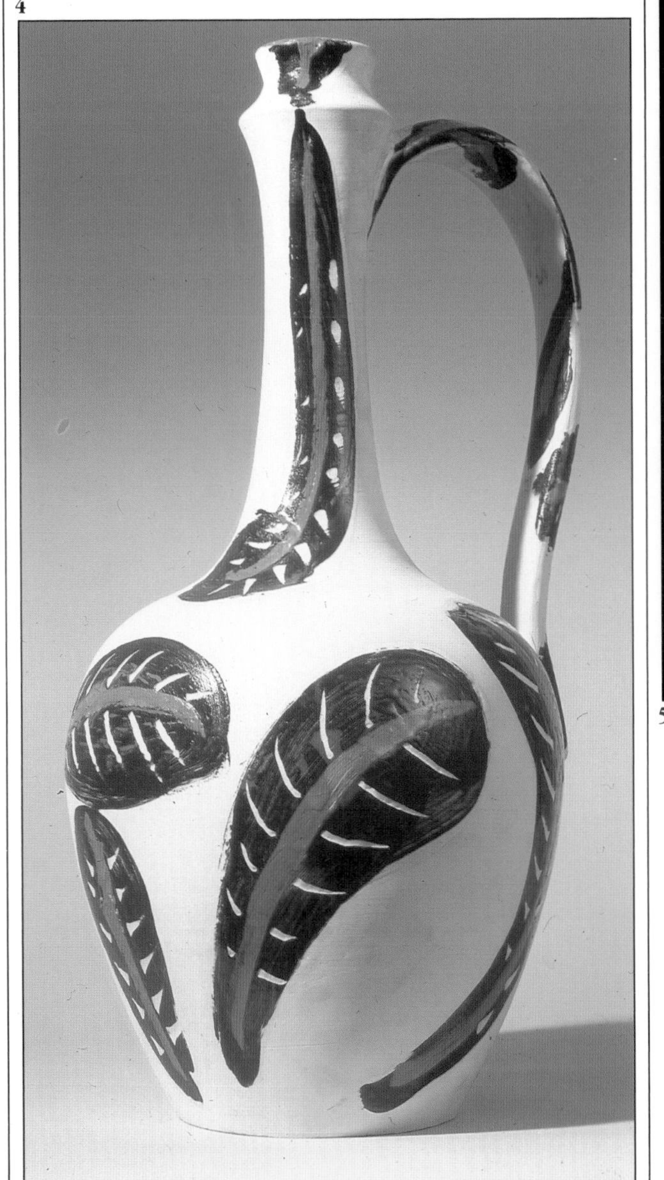

5

Personages, a dish decorated by Joan Miró, c. 1954-56 (**5**). Miró worked with his fellow Catalan Jose Llorens Artigas who was an important potter in his own right. This work was executed using a high temperature technique which they employed between 1953 and 1956. Artigas also worked with Raoul Dufy, Albert Marquet and Georges Braque.

A stoneware vase by Hans Coper.

CHAPTER · EIGHT
THE STUDIO POTTER

"...a pot in order to be good should be a genuine expression of life. It implies sincerity on the part of the potter and truth in the conception and execution of the Work."

BERNARD LEACH – A POTTER'S BOOK, 1940

THE STUDIO POTTER

A stoneware vase by Bernard Leach, c. 1970. Bernard Leach wrote "All my life I have been a courier between East and West" and more than anyone else through his writing and teaching as well as his pottery he brought to the West an understanding of the stoneware traditions of China, Japan and Korea.

Perhaps the most important strand in the development of modern pottery is that of the individual artist potter or studio potter. This cannot be isolated entirely from the Arts & Crafts movement or the work of the industrial designers.

Early influences

The first individual potters to work and experiment with Oriental glazes and Middle Eastern designs were in France; this was largely a reaction to the prevailing industrial ceramics. Théodore Deck is usually credited with being the first of these potters – having worked as a stovemaker, he settled in Paris and established a workshop in 1856 where he sought to emulate Iznik wares. He worked with other painters and created the first European *flambé* glazes, which were exhibited in France in 1884. He later became art director of the Sèvres factory. Ernest Chaplet was the foremost of a very active group of French potters who experimented with glaze effects from the 1870s, and were strongly influenced by Chinese monochromes. The early French pioneers had an important influence on the art pottery of Europe and America.

Bernard Leach studied painting at the Slade School of Fine Art and in 1909 went to Tokyo to teach etching. He was captivated by the Japanese attitude to pottery; they did not regard it as a minor art, but considered it to be a worthy vehicle for the highest aesthetic and spiritual ideas.

He studied under the 6th Kenzan (the Kenzan being the traditional structure of the Japanese potteries) and came to empathize completely with Japanese ideals. He was one of the very few Westerners ever to be totally accepted by his fellow Japanese potters and was granted the right to inherit the title of Kenzan. After travelling to China and Korea, Leach returned to England in 1920 with his pupil Shoji Hamada, and established a pottery at St Ives where they worked with raku, stoneware and traditional English slipware. Leach insisted that there should be no division of labour within a workshop and that the potter should perform all the necessary jobs, from the most menial to the creative. His intention was to produce simple domestic wares as much as works of art.

The influence of Bernard Leach is hard to overestimate and *The Potter's Book*, published in 1940, became the bible for a generation of potters. Amongst his many pupils were Michael Cardew, who later taught in Nigeria, Katherine Pleydell-Bouverie and Norah Braden, all of whom were to experiment and spread his ideals. The work of William Staite Murray who had been potting before World War I, was rather overshadowed by that of Leach, but he had sought to raise pottery to a fine art. His work was also in stoneware with strong Oriental influence; later it was much influenced by Shoji Hamada.

A stoneware dish by Shoji Hamada. After working with Bernard Leach in England, Hamada returned to Japan where he was instrumental in the Mingei, *or folk art movement, which echoed the tea masters' concern with the natural dignity of everyday wares.*

Post-war developments

Since World War II there has been a growing spirit of internationalism in ceramics, stimulated by the emigration of important potters from continental Europe to America and Britain and the widespread dissemination of ceramic magazines and reviews. Potters have challenged the limits of their craft both technically and artistically. Advances in kiln technology and materials have made work easier and have opened new creative avenues. Many modern potters have chosen to hand build their pots rather than throw them on the wheel, as this increases the variety of shapes possible, allows for asymmetric forms and also for more conscious control. Stimulus has come from outside the ceramic traditions, particularly the fine arts, and many potters, notably in America, have sought to break down the barriers between the applied and fine arts. Much ceramic sculpture has really become a branch of the fine arts and, arguably, is no longer part of the ceramic tradition. But conventional ceramic forms, such as vessels (in the widest sense of the word) continue to inspire most potters. Vessels are not necessarily created to be functional, however, and the freedom to experiment is total.

American ceramics

As in the fine arts, America has dominated the modern ceramic scene. Immigrants such as Maija Grotell from Finland, Gertrud and Otto Natzler from Vienna, and the Bauhaus-trained Frans and Marguerite Wildenhain reinforced and invigorated the ceramic tradition already established by pioneers such as Charles Binns. A slightly later arrival, and one of the most influential, was Ruth Duckworth from Germany, who arrived in Chicago in 1964 after spending some years in England.

As well as the European influences brought by these immigrants the East was a source of much inspiration, particularly with regard to Japanese teawares and the ideas of Zen Buddhism.

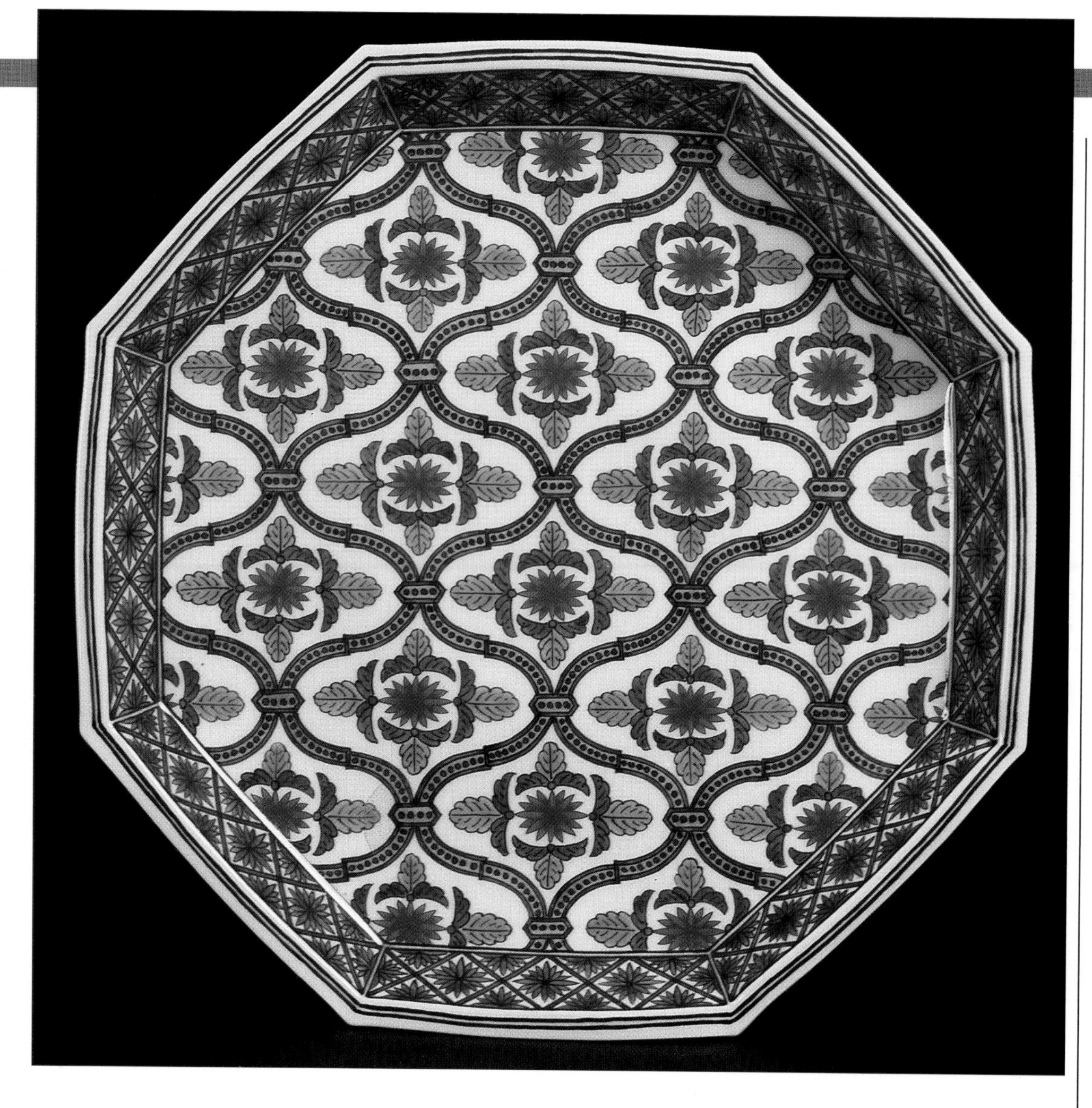

An octagonal porcelain dish, c. 1980 by Imaizumi Imaemon XIII, of the 13th generation of a family of potters who have worked in Arita, Japan, since the 17th century. This dish recalls some of the features of Nabeshima porcelain with echoes of Turkish Iznik wares.

Probably the most influential of the American potters has been Peter Voulkous, who taught at the Otis Art Institute in Los Angeles and in 1959 went on to the University of California at Berkeley. Voulkous drew inspiration from a wide range of sources: the Orient, artists in ceramics like Miró and Picasso, the European tradition and particularly contemporary painting and sculpture. Voulkous' work exhibits a high standard of craftsmanship whilst often giving an expressionistic appearance of roughness and spontaneity. He has done much sculptural work and was instrumental in breaking down the barriers between ceramics and fine art.

American potters have felt less constricted by convention than their European counterparts. Although there are many fine potters, such as Rudy Staffel, who are creating new interpretations of traditional forms, movements such as the West Coast "Funk Ceramics" have drawn on Pop Art to challenge, entertain and shock – the works of Robert Arneson and David Gilhooly exemplify this approach.

English and European pottery

England was also the beneficiary of emigration. Two refugees from Nazism, Hans Coper from Germany, and Lucie Rie from Vienna, have had more influence than any other potters on the development of British ceramics since the war. The monumental sculptural qualities of Coper's work and the beautifully potted bowls of Rie, have helped to free potters from the rather overpowering Anglo-Oriental tradition of Bernard Leach (although this tradition still has many fine exponents, such as David Leach and Richard Batterham). Amongst the more recent potters, the geometrically painted vessels of Elizabeth Fritsch are pre-eminent. Britain has never had so many excellent potters working in such a wide range of

"Dégourdi", c. 1986 by France Martin, hard-paste porcelain. The French tradition of studio pottery dates back to the middle of the 19th century. The proliferation of ceramic publications has tended to break down the barriers between national traditions.

styles. The finely crafted salt glaze stonewares of Wally Keeler have reinterpreted traditional forms. There are the brightly coloured, asymmetric and non-functional vessels of Alison Britton and the mysterious sculptural work of Jacqui Poncelet, to name a few.

The great variety of potters at work in America, Europe and Australia makes it dangerous to generalize about national styles but certain attitudes and traditions do exist. The Dutch and Scandinavians are largely concerned with design qualities rather than the craft aspects of pottery. Many German potters have concentrated on glaze qualities; they do not like eccentric shapes and prefer geometric forms of decoration. The quality of German craftsmanship is often of a very high order, as can be seen in the work of Karl and Ursula Scheid, Hans and Renate Heckman and Gotlind Weigel.

France, like Germany, has tended to go its own way and has benefited from the close links maintained between

"Antiquey Mug" by Anne Kraus, USA, 1989. In recent years the solemnity of much studio pottery has given way to more personal and sometimes humorous work.

potters and fine artists. Amongst the many fine potters, particularly notable are Claude Varlan, whose work is strongly influenced by Japanese stonewares, and the superbly finished and burnished wares of Pierre Bayle.

The Memphis Design Group of Italy, under the leadership of Ettore Sottsass, has really grown out of industrial design concepts rather than ceramic traditions and has launched a frontal attack on functionalism and modernism, the legacies of the Bauhaus. The ceramics of Matteo Thun, an architect and designer, and the craftsman Alessio Sari, keep the appearance of craft to a minimum and use bold shapes with bright colours to create works of great humour. Echoes of Art Deco and a cheerful disregard for accepted taste have, at best, created striking and arresting wares.

The East

Modern Japanese potters continue to develop their ceramic tradition, producing work of the highest quality which can command huge prices and is rarely seen outside Japan. It is not only the folk traditions that flourish but also fine porcelains and Chinese-influenced wares. In the early 1950s the avant-garde Sodeisha group was formed, whose influence continues in strong sculptural vessels.

The strong ceramic tradition of Korea persists and the revival of the Chinese ceramic industry holds great promise for the future. Potters working in the Yixing stonewares are already producing work to rival that of their great past.

The enormous versatility of modern ceramics as a medium for artistic expression seems set to ensure it a very lively future, and the public appreciation of pottery has never been higher.

INFLUENCES

*An English medieval jug (**1**), 13th/14th century from Coventry and a slipware coronation dish (**2**) by William Talor, c. 1675. The strength of medieval forms and the spontaneity of slipware decoration appealed particularly to the early studio potters.*

*A Chinese Henan black-glazed jar (**3**), Northern Song dynasty, 12th century. The Chinese stoneware tradition produced some of the most handsome of all pots which have served as models for many studio potters.*

3

2

1

Victor Vasarely, Supernouvae *(**5**), 1959–61. As the dividing line between the fine and applied arts has become blurred potters have gained inspiration from movements in contemporary art. Abstract decoration lends itself well to pottery and features of Op Art can be seen on recent pottery.*

*A Chinese Guan ware vase (**4**), Southern Song dynasty, 12th/13th century. The decoration relies entirely on the quality of the glaze and the intentional broad crackle. The Chinese potters of the Song dynasty developed a superb range of monochrome glazes that still form the basis of much modern ceramic work.*

4

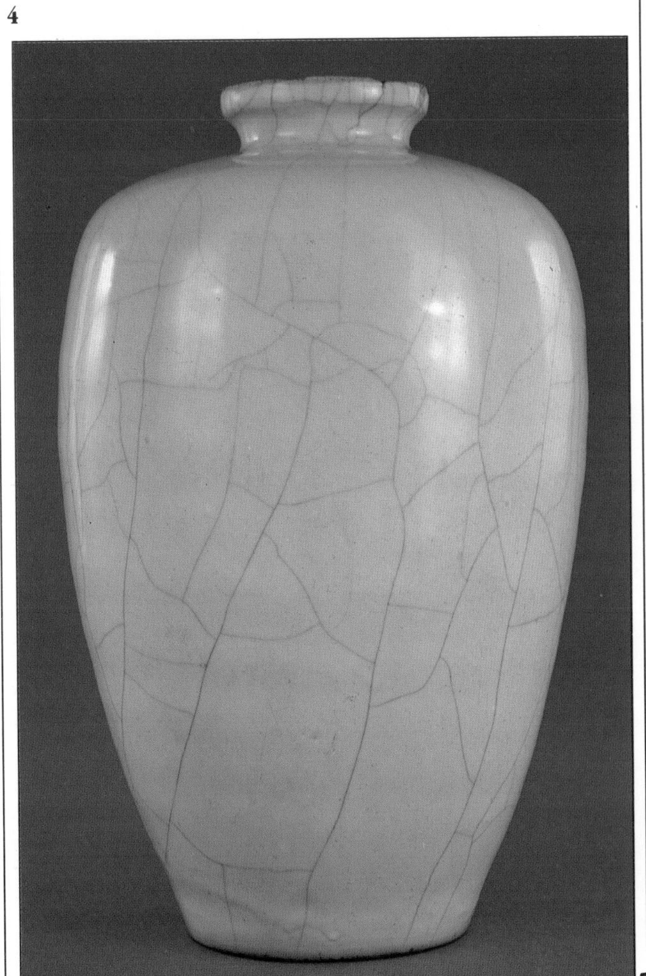

5

EARLY STUDIO POTTERS

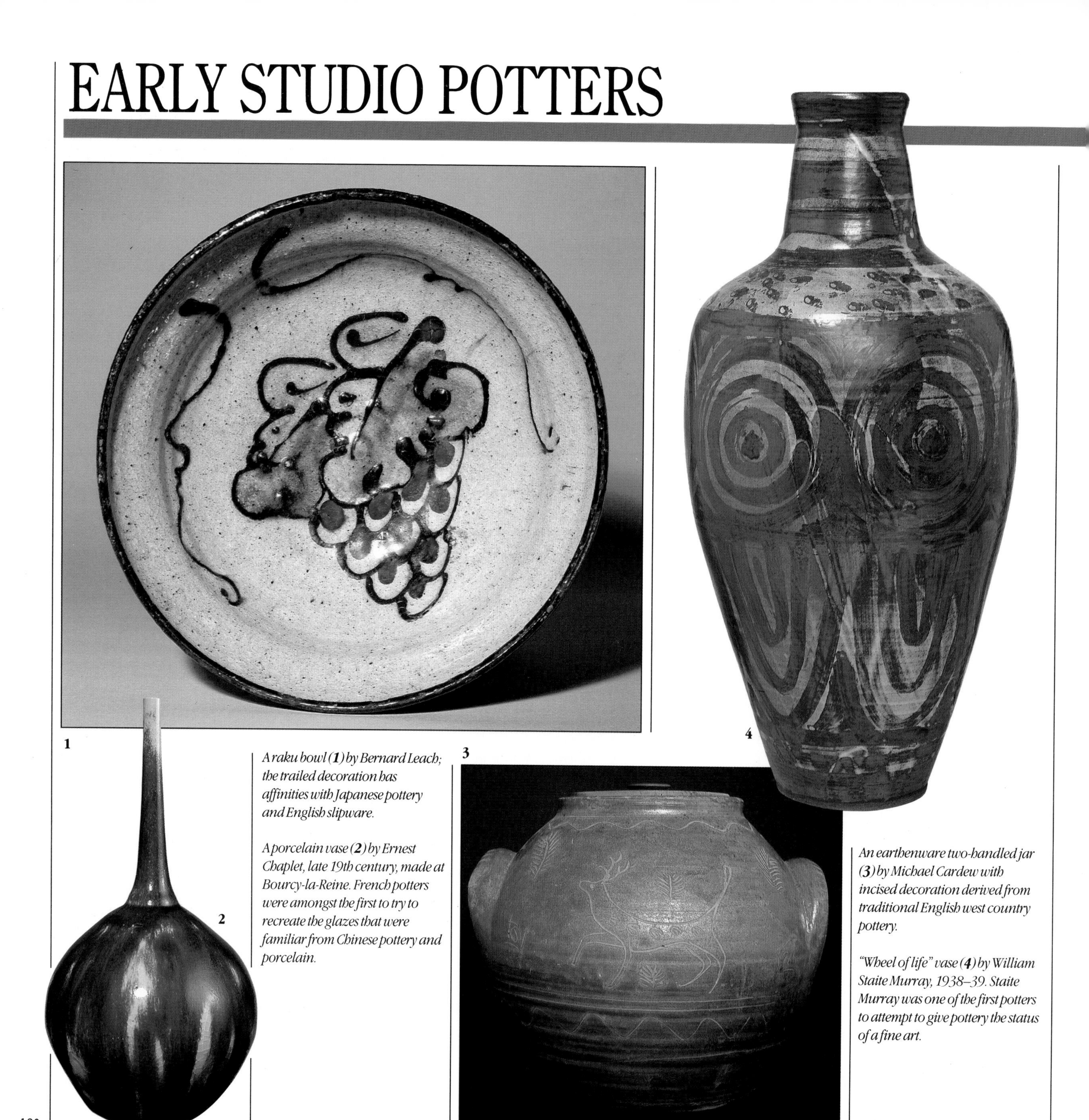

A raku bowl (**1**) by Bernard Leach; the trailed decoration has affinities with Japanese pottery and English slipware.

A porcelain vase (**2**) by Ernest Chaplet, late 19th century, made at Bourcy-la-Reine. French potters were amongst the first to try to recreate the glazes that were familiar from Chinese pottery and porcelain.

An earthenware two-handled jar (**3**) by Michael Cardew with incised decoration derived from traditional English west country pottery.

"Wheel of life" vase (**4**) by William Staite Murray, 1938–39. Staite Murray was one of the first potters to attempt to give pottery the status of a fine art.

*A stoneware "fish" vase (**5**) by Bernard Leach, c. 1930. Painted on a cream-white ground, this combines a Chinese form with the spontaneous painted decoration that Leach perfected in Japan. The muted colours are typical of the subdued approach of the early studio potters.*

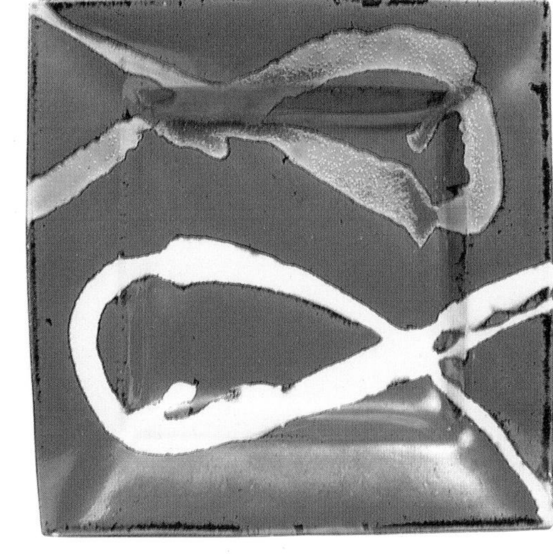

6

*Two bottles and a tray (**6**) by Shoji Hamada. The traditional Japanese forms are decorated with a boldness that is characteristic of Japanese* Mingei *(folk art) pottery.*

5

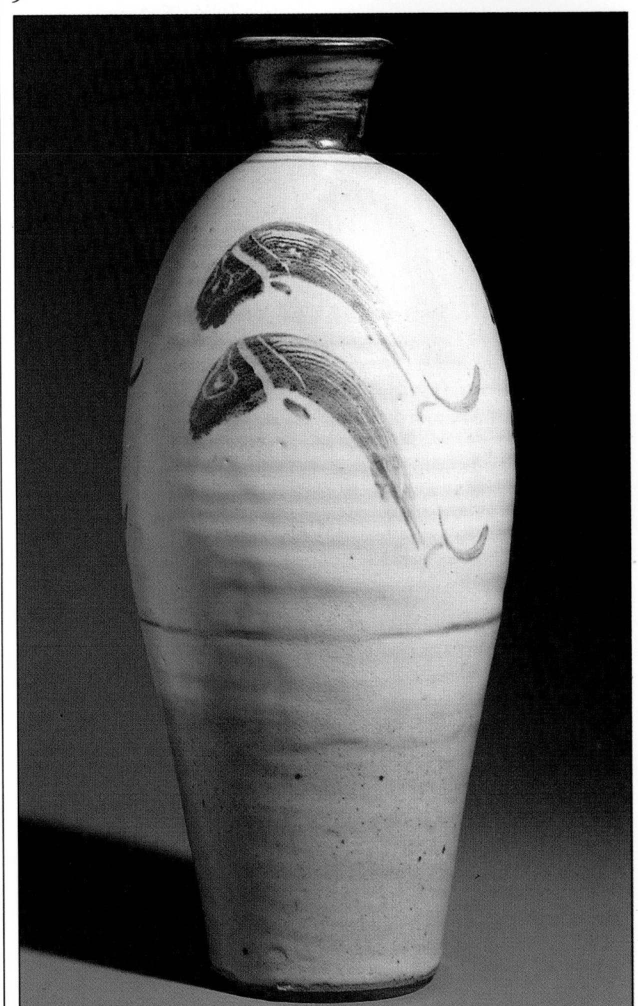

7

8

*A stoneware "bird" charger (**7**) by Michael Cardew. After studying with Bernard Leach, Cardew worked in West Africa and was influenced by the native pottery.*

*A stoneware vase (**8**) by Bernard Leach in the style of the Chinese Henan wares of the 12th and 13th centuries.*

POST-WAR DEVELOPMENTS

1

*A recent vessel by Ruth Duckworth (**1**) 1990. Ruth Duckworth was born in Germany and emigrated to Britain shortly before the Second World War. She introduced a sculptural element into the local tradition before going to Chicago in 1964 where she has continued as one of the most influential potters. The movement of many potters in the post-war years has helped to bridge the gap between national traditions.*

2

*A bowl by Lucie Rie (**2**). Lucie Rie was born in Vienna where she trained under M. Powolny before moving to London in 1938 where she still continues to work. With Hans Coper, with whom she shared a workshop, Lucie Rie has evolved a style in porcelain and stoneware in marked contrast to the oriental-inspired tradition dominated by Bernard Leach.*

3

4

5

*An earthenware bowl with tiger's-eye glaze (**3**) by Gertrud and Otto Natzler, 1958. The Natzlers moved to America from Vienna in 1938. Gertrud threw the pots whilst her husband developed an impressive range of glazes.*

*A stoneware vase (**4**) by Hans Coper, thrown in sections and joined. Coper came to England from Germany in 1938 and his textured forms are amongst the finest pots made in Britain since the war.*

Abstract Plate *(**5**), stoneware, by Peter Voulkos, c. 1963. Perhaps the most influential post-war American potter, he studied painting before changing to ceramics where he was instrumental in breaking down the barriers between sculpture and pottery.*

CONTEMPORARY STUDIO POTTERY

1

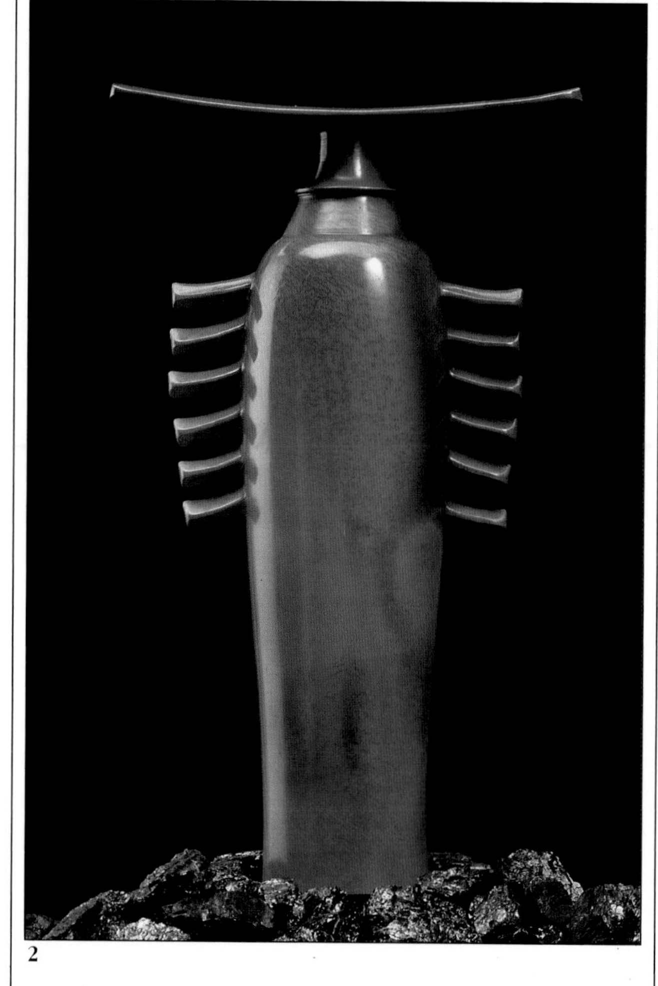

2

*Mamdov Eldar Mursal Ogly (**1**), Azerbaijan;* Object, *1987, glazed stoneware. This boldly decorated and humorous sculpture form is full of references to the cultural traditions of Islam.*

*Pierre Bayle (**2**), France; highly burnished stoneware. This fine pot is finished with considerable care and demonstrates equal attention to the craft and the design.*

*Janet Mansfield (**3**), Australia; salt-glazed stoneware. This potter has been greatly influenced by the Japanese tradition in which the ritual of making the pot is as important as the work itself.*

3

4

5

*Rick Dillingham (**5**), USA; 1986. Dillingham has carefully contrived to give the impression of a broken and recreated pot. The form and decoration is reminiscent of the pottery of the American Pueblo Indians.*

*Nathalie du Pasquier (**6**),* Carrot. *This French-born designer moved to Italy in 1979 where she came into contact with the Memphis group. The solemnity of much 20th century pottery has been challenged with wares that show scant regard for function and have anarchic decoration.*

*Anna Silver (**4**), USA; cobalt blue vase with gold rim, 1988. In recent years the use of bold colours and brilliant glazes has returned to the studio as on this vase using an ancient Mediterranean form.*

6

1

2

3

*Ken Ferguson (**1**), USA; hare teapot, 1989, stoneware. This overscale work is over a foot high and rather heavy which purposefully denies it the function for which is is intended.*

*Ewan Henderson (**2**), UK; mixed stonewares and other clays. This richly textured pot tests the materials of its manufacture almost to the point of collapse in firing.*

*Claude Varlan (**3**), France; a ruralist influenced by Japanese stonewares with their studied randomness and asymmetry.*

4

*Magdalene Odundo (**4**), UK; highly burnished stoneware. The disciplined form contrasts with the irregularities of colour caused by the firing.*

*Edward S. Erbele (**5**), USA;* An influence of two men, Jung and J. Campbell, *1989, porcelain. The complex imagery derives from the potter's subconscious and the psychoanalytical work of Carl Jung.*

*Elizabeth Fritsch (**6**), UK; grogged stoneware with coloured slips. Her painstaking work distorts familiar forms and exhibits carefully contrived geometric decoration.*

5

6

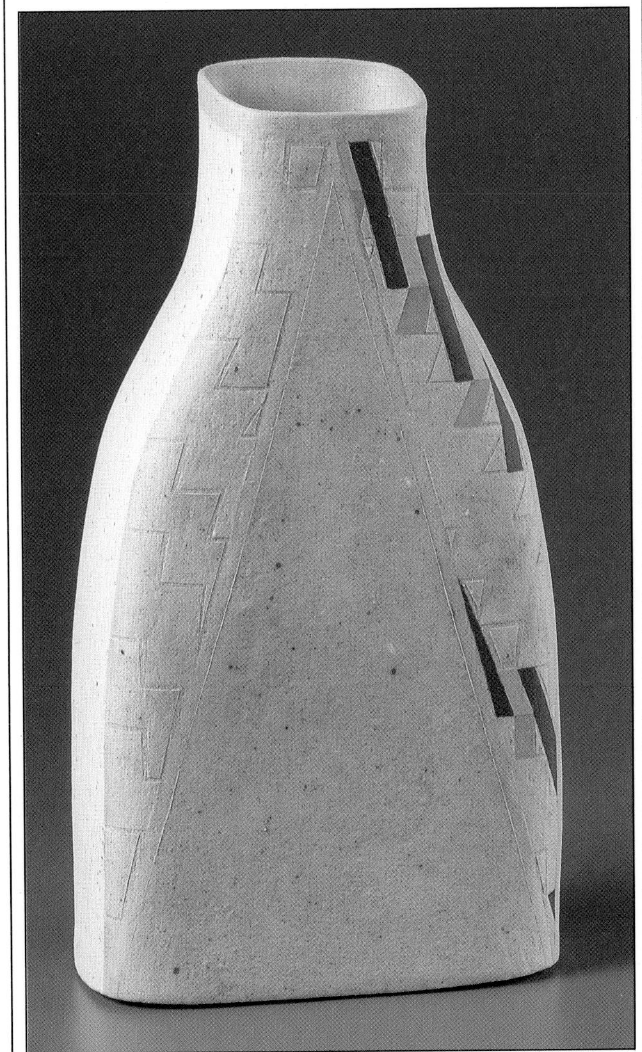

INDEX

Page numbers in *italic* refer to the illustrations and captions

V

W

X

Y

Z

CREDITS

Quarto would like to thank all sources who provided photographs for this book, and for permission to reproduce copyright material. Special thanks are due to the Christie's Colour Library, who provided most of the pictures. The remaining pictures were supplied by the following sources:

Abbreviations:
above a
below b
left l
right r
centre c

Ashmolean Museum
28, 29, 90br, 911, 92ar, 93bl, 98, 99al and b, 111l, 119br, 133al, 145ar.

Pierre Bayle
184ar.

Galerie Besson
181b, 183ar, 187br.

Museum of Fine Arts, Boston
146l.

Bridgeman Art Library
7al, 10, 22a, 23al, 24r, 67b, 71ar, 82m, 90ar, 93a, 102, 103, 106br, 108ar, 113al, 114bl, 116r, 118r, 132bl, 135bl, 156, 158r, 160r, 161a, 166br 170ar, 175, 178l, 179r, 189ar, 181br.

By courtesy of the Trustees of the British Museum
6b, 18, 19, 24l, 25, 26, 27, 37r, 64r, 84, 86, 88al and ar, 90l, 91r, 92b, 93br, 94, 95, 97, 99ar, 100, 110ar.

Garth Clark Gallery, New York
182a, 183al and b, 185l and ar, 186l, 187bl.

Moira Clinch
185br.

Alistair Duncan, New York
157, 162al, 163l.

E T Archive
12, 20, 38ar, 39b, 41bl, 66a, 85, 89bl, 106l, 107r, 119cr, 121ar, 128, 130b, 159al, 179l.

Werner Forman Archive
21, 22b, 23ar, 30, 31.

Barry Friedman Gallery, New York
166, 167.

J. Paul Getty Museum, Malibu
141l.

Ewan Henderson
186ar.

Angelo Hornak
88br, 89ar, 116bl, 130a, 139br, 154, 159bl, 168bl, 170bl.

Janet Mansfield
184br.

Magdaleno Odundo
15.

Mamedor Eldar Ogly
184l.

Percival David Foundation
cover background, 5, 32, 44b, 45cr and br, 46al and r, 47r, 48r, 49l and ar, 51l, 53al, 54l, 55l, 89al.

Musée de Sèvres (photo Hubert Josse)
9, 11, 13, 14, 108l, 109al, 113bl and r, 114r, 116ar, 133bl, 138a and br, 140ar, 147l, 149l, 164r, 177b, 180bl.

Sothebys
6a, 7bl, 13l, 35, 38l and br, 39al, 40l, 41al, 42r, 43l, 44a, 45l and ar, 50, 52l, 53bl, 54r, 55r, 56l, 57b, 66br, 73ar and b, 78r, 79bl, 81b, 87, 92al, 96, 108bl, 110br, 111br, 114al, 115br, 116a, 119l, 120r, 121al and b, 123l, 131l, 134l, 136r, 139bl, 140b, 141r, 144b 145al, 146ar, 148c and r, 150, 151, 161bl, 170b, 178br.

Tokyo National Museum/International Society for Educational Information Inc
60, 66bl, 68, 69, 71al.

Claude Varlan
186br.

Victoria and Albert Museum
37l, 52r, 59l and ar, 62, 69r, 79al, 104, 118l, 124, 140al, 155, 176.

John White
177a.

Whilst every effort has been made to trace and acknowledge all copyright holders, we would like to apologise should any omissions have been made.